Stritzke
Custom Molding of Thermoset Elastomers

Bernie Stritzke

Custom Molding of Thermoset Elastomers

A Comprehensive Approach to Materials, Mold Design, and Processing

HANSER

Hanser Publishers, Munich   Hanser Publications, Cincinnati

The Author:
Bernie Stritzke, N9013 Stringers Bridge Road, EAST TROY, WI 53120, USA

Distributed in the Americas by:
Hanser Publications
414 Walnut Street, Cincinnati, OH 45202 USA
Phone: (800) 950-8977
www.hanserpublications.com

Distributed in all other countries by:
Carl Hanser Verlag
Postfach 86 04 20, 81631 Munich, Germany
Fax: +49 (89) 98 48 09
www.hanser-fachbuch.de

The use of general descriptive names, trademarks, etc., in this publication, even if the former are not especially identified, is not to be taken as a sign that such names, as understood by the Trade Marks and Merchandise Marks Act, may accordingly be used freely by anyone.
While the advice and information in this book are believed to be true and accurate at the date of going to press, neither the authors nor the editors nor the publisher can accept any legal responsibility for any errors or omissions that may be made. The publisher makes no warranty, express or implied, with respect to the material contained herein.

Library of Congress Cataloging-in-Publication Data
Stritzke, Bernie.
Custom molding of thermoset elastomers / Bernie Stritzke.
  p. cm.
Includes bibliographical references and index.
ISBN-13: 978-1-56990-467-1
ISBN-10: 1-56990-467-7
1. Elastomers--Molding. 2. Thermosetting composites. 3. Rubber,
Artificial. I. Title.
TS1925.S86 2009
678'.7--dc22
                    2009017203

Bibliografische Information Der Deutschen Bibliothek
Die Deutsche Bibliothek verzeichnet diese Publikation in der Deutschen Nationalbibliografie;
detaillierte bibliografische Daten sind im Internet über <http://dnb.d-nb.de> abrufbar.

ISBN 978-3-446-41964-3

To keep this book in stock, we used a print-on-demand solution. The quality of the product may differ from the original. The content remains unchanged.

© Carl Hanser Verlag, Munich
unmodified reprint of the 1st edition of 2009
Production Management: Steffen Jörg
Coverconcept: Marc Müller-Bremer, www.rebranding.de, München
Coverdesign: Stephan Rönigk
Typeset: Hilmar Schlegel, Berlin
Printed and bound by BoD – Books on Demand, Norderstedt
Printed in Germany

Acknowledgment

I would like to acknowledge the assistance, encouragement and advice from the following individuals. Most of all, I want to thank my wife, Debby, for her understanding, encouragement, and constructive criticisms.

Martin Benjamin

John Dick

David Fleming

Dr. Tom Hall

David Kurth

Karin Pecararo

Fred Sadr

Brent Schooley

Paul Stoeck

Doug Twing

I would also like to thank the companies and organizations that supplied diagrams, pictures, and advice. I encourage you to contact the sources in the footnotes for more detail of their products.

Preface

This book is intended for the custom molder of thermoset elastomers, who already has a working knowledge of most molding methods. It is not intended to be a book of basics or fundamentals, but rather a comprehensive guide to understanding thermoset elastomer molding by analyzing the manufacturing process from the original part specification through the production molding phase. This book was written with the custom molder in mind, and does not cover manufacturing processes to produce tires, dippings, extrusions, or thermoplastic elastomers. This book should be used to optimize/improve existing processes, develop new cost-effective processes, troubleshoot processes, and assist in thermoset elastomer product design. Although recommended, this book does not need to be read in order. Certain chapters may be skipped and should not influence the understanding of others.

Discussion concentrates on thermoset millable elastomers and LSR (liquid silicone rubber). Emphasis is placed on compression, transfer, and injection molding (including LSR). Hybrids of each molding method, including insulated delivery systems, are discussed. This book describes in detail how flashless transfer, DASM compression, valve-gated cold runner injection, and other molding methods work and what applications are best suited for each method. Thermo-Set Elastomers are referred to as TSE throughout this book.

Many books have been written about thermoset elastomer processing, but chiefly from a chemist's perspective. This book covers basic information pertaining to thermoset elastomer chemistry, but only to the extent needed to effectively understand its interaction during the molding process.

Often TSE molding is put in the same category as plastic injection molding. Emphasis in this book is placed on unique differences in TSE molding as it compares to plastic injection molding.

This book shall be used as a general guide. Each TSE molder has their own unique methods and equipment, and therefore this book stakes no claim into the validity of its contents to work in every environment. This book describes processes, equipment, or tooling that may be covered under various patents. It is the responsibility of the reader to research these claims.

The advice and opinions expressed in this book are believed to be true and accurate at the date of printing. Neither the author, editors, nor publisher accept any legal responsibility for any errors or omissions that may have been made.

Contents

1   Introduction to Thermoset Elastomer Chemistry

This book concentrates on molding and processing of TSE (thermo-set elastomers). More specifically, it will focus on the custom molding industry. Custom molders are typically defined as manufacturing companies that do not sell their goods as an end product, but rather are suppliers to end product manufacturers. Therefore, the manufacturing of tires will not be discussed. Furthermore, because TSEs come in a variety of forms and are subject to an assortment of manufacturing processes, extrusion and any manufacturing form other than molding will not be covered. Molding TSEs involves complicated interactions between material, equipment, and people. Successful TSE molders place a great deal of importance on each category and the influences that they have on each other.

To bring clarity to TSE molding, a look at its chemistry is essential. This book attempts to presents TSE chemistry to someone who may not have a chemistry background. Polymer science can be daunting and since many people reading this book may have limited chemistry background, chemistry will be discussed generally, in easy to understand terminology, yet detailed enough to gain a good understanding.

Rubbers, or elastomers, are part of a larger family of materials called polymers (Fig. 1.1).

1.1   Chemistry Overview

A **polymer** (from Greek πολὐ-ς /poli/ much, many and μὲρος /meros/ part) is a large molecule (macromolecule) composed of repeating structural units.

A molecule is a series of atoms that are held together by electromagnetic forces (valence). The molecule to the left (in Fig. 1.2) is a series of hydrogen and carbon atoms that satisfy their valence protocols and form isoprene. There are roughly 100 identified atoms, each having a valence protocol, or a specific number of bond sites. For example, the isoprene molecule to the left has several carbon atoms. A carbon atom has four bond sites. Notice the furthest left carbon atom has a single bond to each of the hydrogen atoms (above and below), and one double bond to the carbon atom to the right. This carbon atom has satisfied its valence protocol and is therefore no longer actively seeking to bond to another atom. Similarly, the valence protocol for hydrogen is one. Notice that each hydrogen atom has only one single bond [1].

Valence protocols can be identified in more common molecules as in Fig. 1.3. Water, or H_2O in the left example, has one oxygen atom which has a valence protocol of two, hence two single bonds to hydrogen (again, each having one bond site). Pure oxygen, or O_2 in the example to the right, has one double bond between two oxygen atoms.

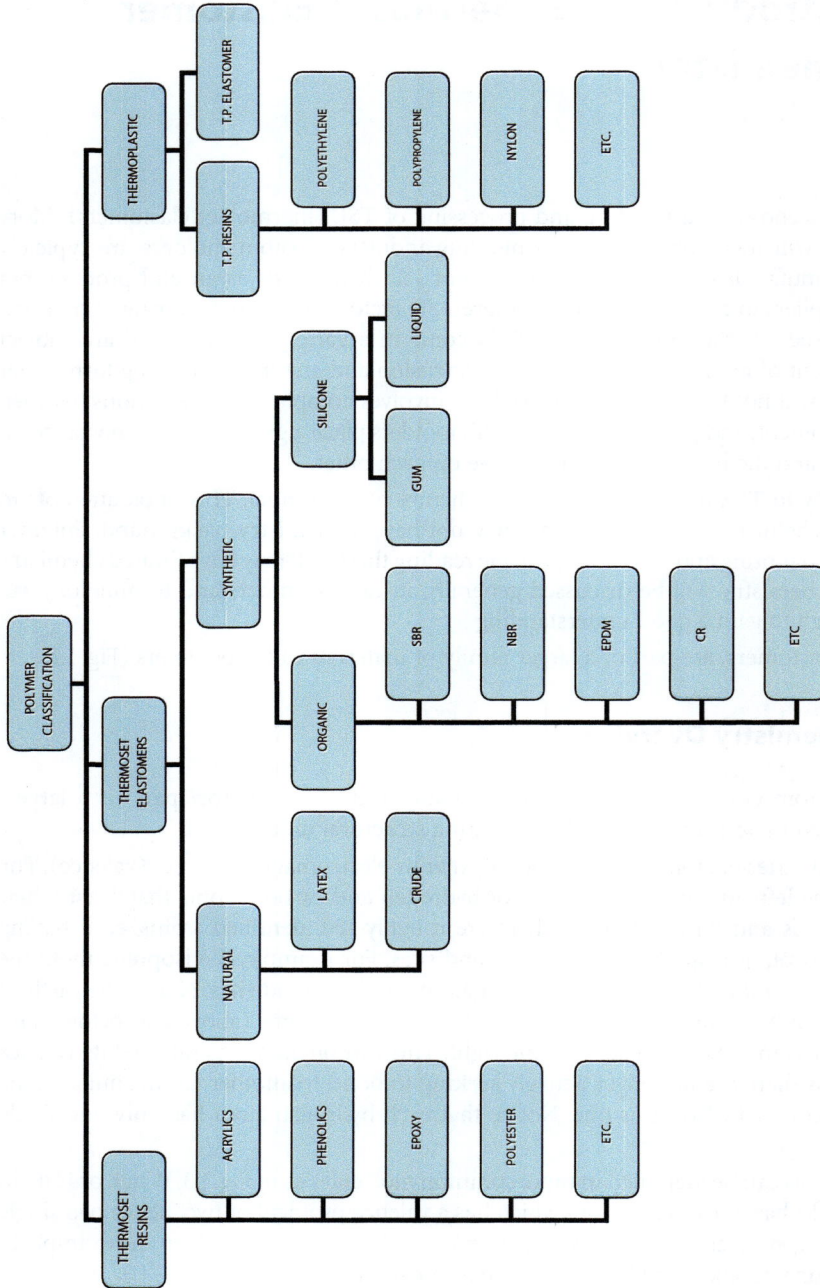

Figure 1.1 Polymer classifications

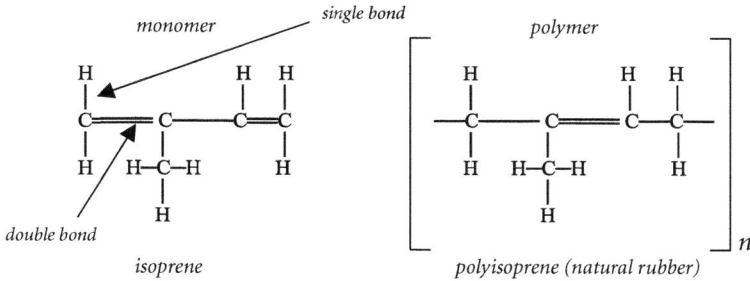

Figure 1.2 Chemical composition of polyisoprene

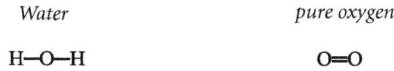

Water   *pure oxygen*

H—O—H   O=O

Figure 1.3 Common chemical compositions

1.2   Polymerization

The chemical structure in Fig. 1.4 demonstrates the polymerization of common plastic poly-ethylene from ethylene gas.

Polymerization subjects the monomer to an energy and/or chemical catalyst. Broadly stated, polymerization is either: *addition polymerization,* which combines monomers without adding or losing any atoms in the process (such as polyethylene), or *condensation polymerization,* which may lose atoms in the form of byproducts. In the case of polyethylene, the polymeriza-tion process breaks the double bond between the carbon atoms rendering the molecule active, or seeking to bond to additional atoms [2]. Each carbon atom now has an active site, and as a result, has an affection for a neighboring carbon atom linking multiple monomers into the resultant polymer. The length of the molecular chain can, to some extent, be controlled during the polymerization process. The longer the polymer chain, the higher the polymer's molecular weight [3]. Polyethylene can be obtained in the form of HDPE (high density poly-

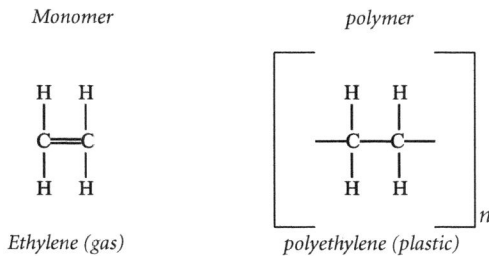

Figure 1.4 Chemical composition of polyethylene

ethylene), or as LDPE (low density polyethylene). This simply refers to a long molecular polymer chain versus a short chain, respectively.

It should be noted that in addition to the strong *interatomic* (valence) forces holding the molecular chain together, there exist much weaker forces holding the molecular chains in close proximity to each other, referred to as *intermolecular* forces [4]. It is said, that as a polymer ruptures, these weaker intermolecular forces are overcome and initiate the break. Polymer molecular chains generally reach into the tens of thousands of repeating units (represented by *n* in the examples above). These chains are, for the most part, a random tangled, twisted, and intertwined nest of strong end-to-end (with occasional off-branches) molecules, held in proximity by the weaker intermolecular forces. Using manufacturing processes such as injection molding or extrusion can intentionally, or unintentionally, minimize the random nature of these molecules. The direction of flow will generally influence the molecules to orient in the said direction. Tests reveal that the resultant product will demonstrate higher tensile strength in the direction of flow, favoring interatomic forces and weaker forces perpendicular to the flow (intermolecular forces).

1.3   Thermoplastic Polymers

Most commercially available plastics fall into this category. In simplest terms, molding thermoplastic polymers is melting and then solidifying the polymer into a desired shape. An obvious advantage that thermoplastic polymers have over thermoset polymers is that thermoplastics can be reused (melted and resolidified). Another advantage is that the thermoplastic molding process is generally much faster than the one for thermosets. The freezing (solidifying) of thermoplastics is much quicker than the chemical crosslink required in thermosets. Thermoplastics process much easier than thermosets. Once a thermoplastic melts, its viscosity is dramatically reduced, and becomes fairly predictable (even considering amorphous versus crystalline materials).

Thermoplastic polymers also include elastomers which are often referred to as TPEs (thermoplastic elastomers). Generally speaking, TPEs have inferior physical properties as compared to thermoset elastomers, particularly at elevated temperatures and/or in harsh chemical environments. Also, since TPEs in large part contain plastic, their hardness range is generally higher than TSEs.

Chemically TPEs can be categorized into several types, but all have at least two distinct characteristics: a hard, plastic-like block (or polymer chain) and a soft, rubber-like block. In essence, TPEs can be thought of as having a thermoplastic binder that can melt, solidify, and be remelted. Attached (linked) to this thermoplastic binder is a soft polymer which can be a thermoset crosslinked rubber, thermoset uncrosslinked rubber, or a softer thermoplastic polymer [5]. In any event, the TPEs' temperature range is determined by the thermoplastic binders melting, or softening, point.

TPEs are molded in much the same fashion as all other thermoplastics. They are purchased in pellet form and fed into an injection hopper, or extruder, melted and then solidified into the desired state.

1.4 Thermoset Polymers

Crosslinking, or vulcanization, is a chemical transformation at periodic locations along the backbone of the polymer chain to create an active site where vulcanizing agents can adhere. There are several crosslinking methods; the most common utilize sulfur, peroxide, or platinum as curing agents. Heat is most commonly used to excite, or drive off, the curing agents. As in Fig. 1.2 of the polyisoprene polymer, the double bond between the two carbon atoms is attacked, creating an active site to which sulfur adheres at every few hundred molecules. Accelerators and/or retarders are used in conjunction with the curing agents to pinpoint the temperature at which most crosslinking initiates [6].

Imagine cutting someone's hair that is 24 inches long. The hair will fall to the ground and assume the shape of the surface it falls onto. Even if you try to squeeze it together, it will have no structure. The hair can be picked up and strands can easily be removed from the pile. Now sprinkle a few dozen burrs into the hair and start kneading the mixture together. The bundle has structure and is difficult to separate. Also, if you push your finger into the bundle, it will resist motion. Once you pull your finger out, it will attempt to retain its original shape. The hair in this depiction is likened to polymer chains. The burrs can be compared to the crosslinks that occur during the cure of TSE.

Albeit very slowly, crosslinking occurs even at room temperature. Some materials are more susceptible to shelf life constraints than others. Nevertheless, viscosity of thermosets are not only nonlinear, but they are non-repeatable — meaning, as temperature rises, crosslinking occurs at an increasing rate, and cannot be brought back to a given viscosity once it has started.

The crosslinking process is regarded as a slow process. When considering a molding process, as thermoset polymers are forced into a mold cavity, it continues to flow, or creep, into small fissures until crosslinking is complete. Thermoplastics, on the other hand can freeze almost instantaneously (at least on the outer surface), stopping the flow. In comparison, solidifying plastic can take seconds, while crosslinking thermosets can take minutes. Couple this with the fact that thermoset molding is typically performed at 300–400 °F, thermoset materials continue to heat up as they cure in the mold. Laws of physics tell us that as solids heat up they expand, and as they cool they shrink. In thermoset molding, thermal expansion generates tremendous internal pressures and the displaced material forces its way through parting lines in the mold cavity. Conversely, plastics enter the mold hot and are cooled — shrinking the molded part. Plastics do not expand in the cavity and as a result are not as susceptible to flash. This, in large part, is why many thermosets require secondary operations to deflash the end product. It should be noted that in thermoset molding, gasses are often a result of the crosslinking process. Vacuum and adequate venting (and frequent mold cleaning) is essential to good molding practices.

1.5 Organic and Silicone Elastomers

Thermoset elastomers are categorized into natural or synthetic classifications. Natural rubber is produced from the latex of the Hevea brasiliensis plant (rubber tree) indigenous to South

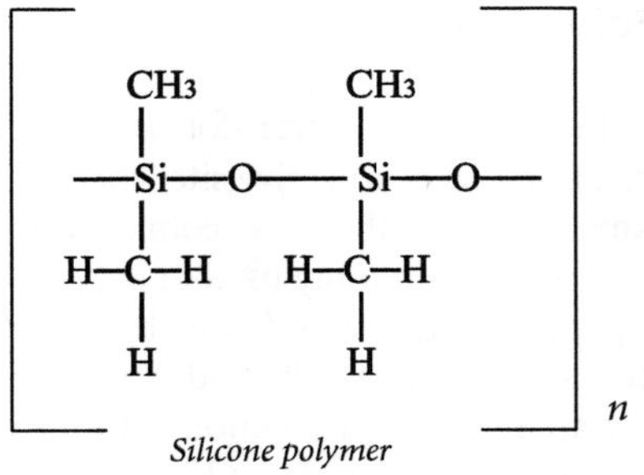

Figure 1.5 Silicone polymer

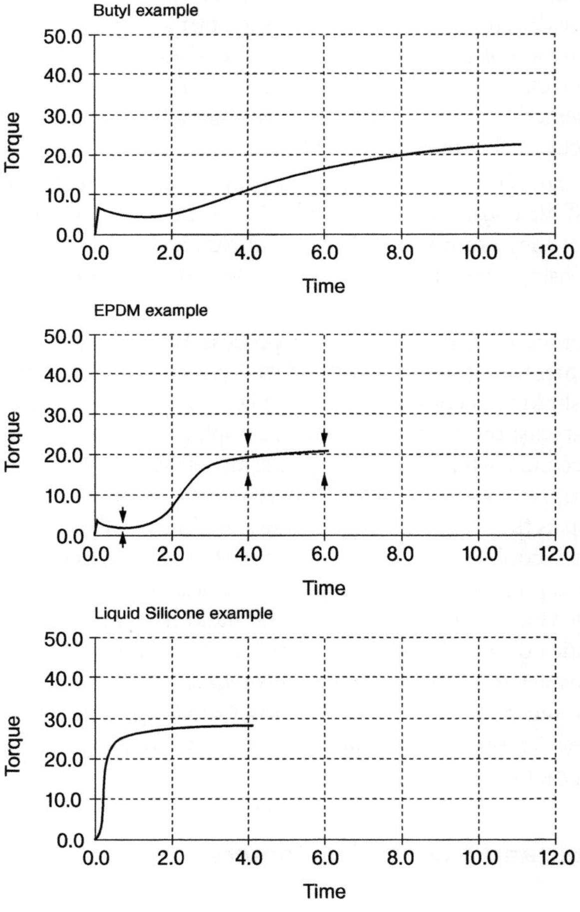

Figure 1.6 Various cure curves

America, particularly the climate and elevation of Malaysia. Natural rubber is chemically known as cis-1,4 polyisoprene (it should be noted that the polyisoprene family of polymers has both natural and synthetic versions) [7]. Synthetic elastomers are categorized into either organic or silicone. Organic polymers have a molecular backbone that consists of carbon-to-carbon linked atoms, as shown in previous examples, and for the sake of our discussion, are petroleum-based.

Silicone, on the other hand has a silicon-to-oxygen backbone (see Fig. 1.5) that gives silicone an inherent high temperature range (this same chemical bond is found in high temperature resistant materials such as quartz and sand) [8].

Double carbon bonds, which may contain a degree of unsaturation, and are found in many organic compounds, are susceptible to ozone attack. Silicone has no double bonds and therefore displays excellent weather and ozone resistance.

1.6   Cure Rates

Rheometer tests are performed on three different compounds at the same temperature in Fig. 1.6. Rheometer testing will be covered in detail in other chapters, but the charts demonstrate the rate of cure of various TSEs. As the curve begins to flatten, the TSE is said to be cured. Notice the curve for butyl shows a cure of more than 8 minutes, while liquid silicone cures in less than one minute.

1.7   Conclusion

A brief background into the chemistry of TSEs is helpful in developing molding processes. With basic knowledge of the interactions between materials and processes it is easier to troubleshoot processes.

References

1. Morton, M., *Introduction to Rubber Technology*, Van Nostrand Reinhold Co., NY (1969), p. 45–46.

2. Morton, M., *Introduction to Rubber Technology*, Van Nostrand Reinhold Co., NY (1969), p. 49–51.

3. Sommer, J. G., *Elastomer Molding Technology: a comprehensive and unified approach to materials, methods, and mold design for elastomers*, Elastech, Hudson, OH (2003), p. 15.

4. Morton, M., *Introduction to Rubber Technology*, Van Nostrand Reinhold Co., NY (1969), p. 51–52.

5. Dick, J. S., *Rubber Technology: compounding and testing for performance*, Carl Hanser Verlag, Munich (2001), p. 264–268.

6. Long, H., *Basic Compounding and Processing of Rubber*, Rubber Division, American Chemical Society, Inc., (1985), p. 156–158.

7. Semegen, S. T., Fah, C. S., Natural Rubber, *The Vanbderbilt Rubber Handbook* (1978), p. 18–35.

8. Noble M. G., Silicone Elastomers, *The Vanderbilt Rubber Handbook* (1978), p. 216–217.

2 Compounding, Mixing and Equipment

2.1 Introduction

This chapter describes the mixing of common TSE types: natural and synthetic gums. Liquid silicones are described in more detail in their own chapter. The intention is to have the reader gain an understanding of how TSE is compounded, mixed, and how the mix can influence the process downstream. Although not specifically necessary, many custom molders have compounding and mixing capabilities in-house. Large mixing houses have technical resources available to custom compound and can typically mix with larger and more efficient equipment to offer less expensive materials.

2.2 Compounding

Compounding is the creation of a TSE formulation that:

- Satisfies the end product's physical and chemical properties
- Can be adequately mixed and handled with available equipment throughout the manufacturing process
- Is optimized to the intended processing method

Most large TSE molding companies have chemists on staff seasoned with years of experience in developing TSE compounds. This is partly due to the fact that these same large companies, more often than not, have their own TSE mixing capabilities. Even if a company cannot justify capital-intensive mixers, they may have mills to add ingredients on-site to avoid shelf-life constraints, or to shape into preps for a specific manufacturing process. Furthermore, most TSE molding companies, regardless of size, will have lab equipment and test equipment to perform some compounding in-house.

As will be described in later chapters, industry standards for presses, molds and processes for molding TSE are far fewer than what exists in the plastic injection molding industry. For this reason, each TSE molding plant seems to have its own compounds tailored to their plant's specific needs. What makes matters worse is that compound development is seldom shared outside the plant. The TSE industry is stereotyped as "only invented here," or "my special ingredient," which translates to the industry as "compounding can be thought of as more of an art than a science".

Contrast this with the plastic injection molding industry which purchases off-the-shelf plastics that suit a variety of needs and processes. Most plastic materials can be categorized as general purpose, or serve a variety of applications. Unless they purchase in huge quantities, the plastic injection molders have little influence on the material composition.

Generic Formulation for Nitrile (NBR) Compound

Chemicals	HPR
NBR Polymer	100.00
Carbon Black	50.00
HISIL	35.00
Stearic Acid	1.00
Zinc Oxide	5.00
Antioxidant	2.00
Plasticizer #1	10.00
Plasticizer #2	5.00
Sulfur	0.50
Accelerator #1	1.00
Accelerator #2	2.00
TOTAL	**211.50**

Generic Formulation for Gum Silicone Compound

Chemicals	HPR
Silicone Polymer #1	60.00
Silicone Polymer #2	40.00
Color Pigment	1.00
Peroxide	2.00
TOTAL	**103.00**

Figure 2.1 Typical TSE formulations [1]

Typical TSE compounds include the following categories of chemicals:

- Base polymer
- Crosslinking agents
- Accelerators
- Activators/retarders
- Antidegradants
- Processing aids
- Fillers
- Plasticizers
- Color pigments
- Special purpose ingredients

2.3   Mixing

Most TSE molding compounds are mixed in an internal mixer. There are many types of internal mixers, but all follow the same basic principle. Materials are fed through a hopper at the top of the machine, which leads to rotors that mix the material. Once the materials

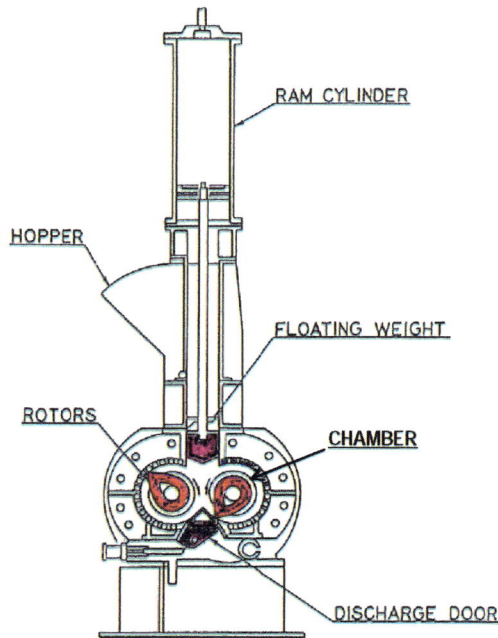

Figure 2.2 Cross-section of mixer [2] (Courtesy of Kobelco Stewart Bolling)

are added, the mixer is closed to avoid loss of mix or contamination. A ram pressurizes the mix from the top. Additional ingredients are added throughout the mixing process. Once the mixing is complete, a door opens at the bottom of the mixer (or a portion of the machine tilts) and the resultant mix is dumped onto an open mill below. Depending on the size of the mixer, a batch can weigh up to 1000 lbs. Typical mix times exceed 2 minutes.

The mixing of TSE offers a great deal of potential variation. In fact, many in the industry would suggest that the greatest variable in TSE molding lies with the inconsistencies of the TSE mix. Variations exist with the raw polymer, as well as the oils and powders that are mixed with it. Couple this active mix with heat and shear stress and a mixed batch can easily become scorched, or lack dispersion.

Once a batch of TSE completes the internal mixer cycle, it needs to be cooled quickly. These batches could weigh several hundred pounds and are a chemically active mass several feet thick. Left alone, this mass of material will self-insulate and cause heat inside to initiate crosslinking. Therefore, after leaving the internal mixer, the batch is dropped onto a mill. The mill has two large rollers which turn in opposite directions. Once the TSE hits the mill, the material is squeezed through the nip — or gap — between the two rollers. This thinning of the mix allows cooling penetration to avoid premature crosslinking. The material is then stripped off of the mill. Further cooling can be accelerated by adding water.

Figure 2.3 Side view of mixer [2] (Courtesy of Kobelco Stewart Bolling)

Figure 2.4 Rotors [2] (Courtesy of Kobelco Stewart Bolling)

Shear heat is generated during the mixing process. Extended exposure to this heat can ruin a batch. For various reasons, many materials require a two-pass mix. This means that the entire mixing operation cannot be accomplished the first time around. The batch needs to be dumped, cooled, and remixed to add more ingredients.

Consider two mixed batches of the same compound. Each batch contains perfectly identical percentages of each ingredient; each ingredient has perfectly identical physical properties,

Figure 2.5 Tilt mixer [2] (Courtesy of Kobelco Stewart Bolling)

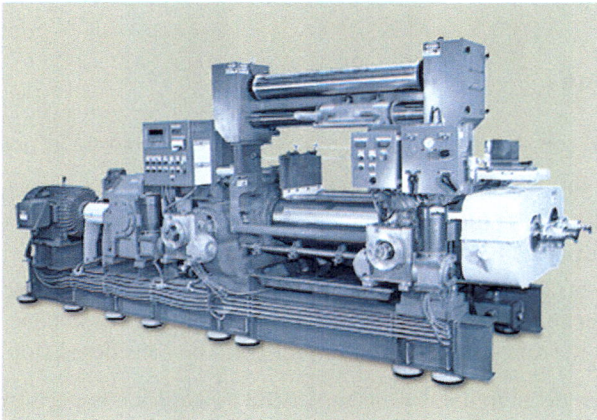

Figure 2.6 Mill [2] (Courtesy of Kobelco Stewart Bolling)

yet each mixed batch can have dramatically different physical properties. Temperature and timing of the mixing process can have a greater impact on the variation of the mixed TSE than the actual variation in the ingredients themselves. A few extra seconds in the mixer, or a delay in getting the batch stripped on the mill can impact the batch. Even the beginning of the batch compared to the end of the batch will offer different results. Consider a several hundred-pound batch of TSE dumped onto an open mill at 230 °F. At this temperature crosslinking is taking place at a slow rate. It can take ten minutes to reduce a TSE batch that is several feet thick into a thin strip to be cooled. The first portion of the strip will have the advantage of cooling immediately, while the balance of the strip will remain in a large mass, and then remain on the mill for several minutes before it reaches a cooling stage.

Therefore, proper controls in the mixing operation are paramount to the success of the molder downstream. Typical controls during the mix are time, temperature, rotor speed, sequence and weight of each ingredient. The amount of energy needed to turn the rotors of the mixer can be an indication of the material's viscosity. Energy levels can also be used to indicate out-of-control mixing.

2.3.1   TSE Compound Batch Release Tests

Rheology tests for compound batch release testing have general acceptance in the TSE industry. In addition to rheology, physical properties are specified for compound certification based on specific requirements of the customer. The following is an example of physical property requirements for a polyisoprene compound [1]:

Tensile = 2500 PSI (Minimum)

Elongation = 770 % (Minimum)

100 % Modulus = 85 PSI (Minimum) – 115 PSI (Maximum)

Hardness (Shore A) = 35 ± 5

Tear Die-B = 250 PPI

2.4   Silicone

For the purpose of this discussion, silicone is available for molding in two forms.

- LSR, or Liquid Silicone TSE (sometimes referred to as LIM — Liquid Injection Molding) is a very low viscosity (pumpable) silicone — meaning, the state in which the material is stored and transported has a pourable consistency.
- Gum silicones, on the other hand, have a thick dough-like consistency and use mixing processes similar to those of an internal mixer (although gum silicone mixers have some uniqueness, for the sake of this discussion they will be considered similar enough).

LSRs are covered in detail in their own chapter, but as a quick comparison, LSR silicones are purchased in drum form, usually in five or 55 gallon containers. No mixing of ingredients is required by the end user other than combining parts A and B, and possibly a color. LSR silicone is ordered in two containers: part A and part B. Both containers have the same silicone, except one will include the cure agent and the other will include chemicals to initiate (excite) the cure agent. The curing agent in LSR is predominantly platinum, which is quickly activated when in contact with a combination of the chemicals to excite the cure and elevated temperatures. For this reason they are housed in separate containers until they combine just prior to the injection chamber of the injection molding machine.

2.5   Conclusion

Compounding and mixing should only be considered as an in-house capability if sufficient expertise exists within the company. In addition, mixing equipment is capital intensive and requires a tremendous amount of space. Unless sufficient volume of material is consumed by a custom molder, or proprietary compounds are feared to become public, mixing should be left to custom mix houses. However, even if outsourced, custom molders should have intimate knowledge of how TSEs are compounded and mixed in order develop and troubleshoot molding processes.

References

1. Sadr, F., Personal interview 4/15/08.
2. Kobelco Stewart Bolling, Inc., Hudson, OH.

3 Materials

This chapter will describe some of the unique differences between available TSE materials along with some important historical achievements in TSE development. Some descriptions of elastomers are taken in their entirety, or abbreviated from referenced sources. Described are only some of the most common materials used for custom molders. For more materials and greater detail, the reader should consider some of the references listed at the end of this chapter or material suppliers.

3.1   Natural Rubber (NR)

"The discovery of rubber dates back to when the Europeans started exploring South and Central America. They observed not only young children playing with balls that bounced, but also crude waterproofing methods for clothing and shoes. The material was called 'cachuc' which means 'weeping wood'. This mysterious elastic and waterproof substance was observed by the explorers as originating from the bark of a certain tree after being punctured. A milky substance secreted and was captured by the natives and poured onto clothing and shoes to form an elastic waterproofing.

The word spread in Europe about this fascinating material prompting all kinds of ideas for its commercial use. An English scientist, Priestley, discovered that this material would rub off pencil marks; he called it 'rubber'. That name stuck with the substance ever since. As interesting as this material was, it had no intrinsic value since the rubber turned stiff and brittle in cold temperatures and soft and sticky at warm temperatures. It was not until 1839 that Charles Goodyear discovered that when the rubber was heated with sulfur, it dramatically changed into a strong, elastic material in cold and warm temperatures. This coupled with Hancock's earlier discoveries on how to shape and process the material finally accelerated rubber into wide-spread use" [1].

Natural rubber, the same material seen by the early European explorers in a bouncy ball is made from the milky colored excretion (latex) from the Hevea Brasiliensis tree. It is the only commercially available non-synthetic rubber. Natural rubber is known chemically as cis-1,4 polyisoprene. Natural rubber is a strong material; its tensile strength makes it the standard bearer in the rubber family. It has a high molecular weight, giving it a high Mooney viscosity. Natural rubber's strength, abrasion resistance, and resistance to deformation make it an excellent candidate for tires and shock isolation devices.

Disadvantages of natural rubber include: poor resistance to oils, fuels, high temperature, ozone, and acids. Often additives and/or blends with synthetic rubbers make up for some of natural rubber's shortcomings.

Natural rubber is considered a difficult to process material. Its high viscosity and high tensile strength resist flow, making injection molding natural rubber difficult. Natural rubber has a long cure time when compared to most other materials. Its high tensile strength makes secondary removal of flash difficult. Natural rubber has an unmistakable odor when molding. It's difficulty to shape and break down in its uncured state (considered green strength) actually offers advantages in molding tires. The green strength allows the uncured rubber to stay in position and be layered with wire bands and other rubber materials.

3.2   Synthetic Polyisoprene (IR) [2]

It was not until the 1950s that the long sought after synthetic version of natural rubber was discovered. It has long been understood that isoprene was the fundamental ingredient to natural rubber, and would be key in developing a synthetic version. With the advent of a new type of catalyst system called "stereospecific", monomer units were able to be selectively joined in a well organized fashion. Initial commercialization of synthetic polyisoprene came in 1960 as Shell Isoprene Rubber, by the Shell Chemical Company. In 1962, Natsyn was introduced by the Chemical Division of Goodyear Tire, and in 1965 Goodrich-Gulf came out with Ameripol SN. These rubbers are all high cis-1,4 content type. There are other synthetic polyisoprene types, but the materials made by cis-1,4 addition best emulate natural rubber.

Not surprising, synthetic polyisoprene is used as a replacement for natural rubber. Synthetic polyisoprene does slightly fall short of natural rubber's physical properties in tensile and modulus, and particularly at elevated temperatures. It also has lower green strength, which may be an advantage or disadvantage depending on the molding process chosen. High green strength is a feature useful in layering materials in tire molding. Synthetic polyisoprene can have a longer cure time than natural rubber.

Synthetic polyisoprene does have advantages over natural rubber. Synthetic polyisoprene is much more consistent than natural rubber. Its lower molecular weight makes it easier to process. Ease and consistency may offset the added cost of synthetic polyisoprene over natural rubber.

Typical applications for synthetic polyisoprene are tires, motor mounts, vibration dampeners, footwear, and rubber bands. Recently, it has gained popularity in the medical industry because it does not contain the potential latex protein allergens found in natural rubber.

3.3   Styrene-Butadiene (SBR) [3]

The copolymer of styrene and butadiene became the most important and widely used synthetic rubber ever produced. But is was not until the 1920s that the first emulsion systems using free radical catalysts to produce high polymerization rates and high molecular weight products were recognized. In the 1930s and the advent of WWII, developed countries wanted to rid their dependence on foreign sources of material. The German government promoted research in the development of synthetic rubber and the first butadiene-styrene copolymer from an emulsion system was produced by the research laboratory of I. G. Farbenindustrie and was known as Buna S. A butadiene-acrylonitrile copolymer (Buna N) followed shortly thereafter.

These first products were of very poor quality compared to natural rubber. However, the technology with considerable improvements and modifications formed the basis for synthetic rubber production in the United States.

As the United States entered WWII, their access to natural rubber became increasingly difficult, while the demand for rubber increased due to the industrial revolution and the need for rubber in military equipment. The US government in association with material companies prompted research on a replacement for natural rubber. In 1942 GR-S (known as SBR today) was launched into production in a government plant. By 1945 the US government financed 15 SBR plants, 16 butadiene plants, and 5 styrene plant which all eventually were sold to the private sector.

Polymerization is done through an emulsion or solution process. In emulsion the predominant ingredients, styrene and butadiene, are emulsified in a water-like solution with emulsifying agents such as soaps. Initial polymerization of SBR was considered a "hot" process and was performed at 41 °C. Later (1947), a "cold" process was developed which used a temperature of 5 °C, and offered superior physical properties over the hotter processed SBR.

The solution process is considered a cleaner process because it does not contain any soap residues common in the emulsion process. These residues can contain as much as 7 % by weight of the polymer bale. This process utilizes a solution of hydrocarbons to produce better stereospecificity. Alkyl-lithium-based catalyst systems are used because they are the only stereo-specific catalysts that copolymerize styrene and butadiene.

Most of the SBR material finds its way to the automotive tire industry. However, SBR is also seen in products like: floor mats, shoe soles, cable insulation, food packaging, adhesives, caulks, military tank pads, and even common chewing gum.

3.4   Polybutadiene (BR) [4]

Polybutadiene is only second to SBR in total synthetic rubber consumption. Nearly 70 % of all BR is used for treads and sidewalls of tires. BR has excellent abrasion resistance and low temperature resistance. In fact, only silicone rubber has a lower temperature resistance. However, BR is difficult to process and exhibits poor wet traction properties and is therefore predominantly used in blends with natural rubber and SBR for tires.

BR is also used as an impact modifier for polystyrene and acrylonitrile-butadiene-styrene plastics. Polymerization is similar to SBR in that the process can be either emulsion or solution based. Typical cure systems for BR compounds are sulfur or peroxide cured.

3.5   Butyl (IIR) [5]

Butyl rubber is a copolymer of isobutylene and a very small amount of isoprene (98 % vs. 2 %, respectively). By far, butyl's greatest property is gas permeability resistance. Coupled with good flex resistance, and resistance to oxidation and ozone, butyl makes an excellent choice for inner tubes, and inner liners in tubeless tires.

Halogenated butyl, known as halobutyl, was developed in the early 1960s, which lead to better processability and allowed for copolymer blends with rubbers such as natural rubber and SBR. This significant event lead to the development of tubeless tires because the addition of the other rubbers allowed the excellent gas permeability resistance of butyl rubber to be bonded to the inner wall of a tire. There are two types of halogenated butyl rubbers: bromobutyl (BIIR) and chlorobutyl (CIIR). Butyl is made by low temperature cationic polymerization.

Other applications for butyl are: vibration dampeners, bladders, steam hose, and pharmaceutical stoppers.

3.6   Ethylene-Propylene-Diene (EPDM)

EPDM is a terpolymer of ethylene, propylene and a small percentage of diene which provides unsaturation in side chains pendant from the saturated backbone [6]. EPDM's most appealing trait is that it has excellent ozone resistance. As such, EPDM is used as roofing material, automotive cooling system seals, wire and cable covers, high voltage boots, and other similar products.

EPDM can be sulfur or peroxide cured. Peroxide cure systems allow for better compression set and a higher temperature range. EPDM has good high temperature resistance, but is prone to attack from hydrocarbons. EPDM's non-polarity gives it excellent electrical resistance properties.

3.7   Nitrile (NBR) [7]

NBR is a highly polar copolymer of butadiene and acrylonitrile. The acrylonitrile content offers NBR its oil resistance that improves as its content increases, but also worsens low temperature flexibility.

NBR has good hydrocarbon resistance, but moderate temperature resistance. As NBR can be cured with a variety of cure systems, peroxide should be used to increase its temperature range. For an even higher heat range, hydrogenated nitrile (HNBR) should be used. The hydrogenation process removes most of the unsaturation in the polymer making it less vulnerable to attack from heat, ozone, and oxygen. HNBR can be made for low, or high temperature resistance.

For improved strength and abrasion resistance, carboxylated nitrile (XNBR or CNBR) is available. For these versions, carboxylic acid groups are added to the polymer chain during the polymerization process. These groups provide additional crosslinking sites during the curing process.

NBR is the least expensive material in the oil resistant rubber category. There is a substantial cost penalty for hydrogenated and carboxylated versions.

3.8   Polyacrylic (ACM)

ACM found favor in the automotive industry because of its outstanding resistance to petro-leum-based oils and fuels. ACM also has good ozone resistance and a high temperature range. However, it has inferior strength, water resistance, and low temperature resistance and as a result was displaced in many applications by ethylene acrylic rubber.

3.9   Ethylene Acrylic (AEM)

AEM is a terpolymer of ethylene, methyl acrylate, and a third monomer (carboxylic acid groups) in a very small amount to serve as a cure site for the resulting polymer. AEM has good high temperature resistance, good oil resistance and moderately low temperature resistance. AEM has replaced much of the ACM used in automotive applications because of better low temperature resistance [8].

Good low temperature properties are derived from its ethylene content, while the methyl acrylate co-monomer provides oil and fluid resistance. The completely saturated nature of the polymer backbone displays excellent resistance to oxidation, ozone, UV radiation and weathering [9].

3.10   Silicone (MQ, VMQ, and PMQ)

Silicones are synthetic rubbers but are derived from organic and inorganic materials. The element silicon, which does not exist naturally, is present in sufficient amounts to positively affect the material's physical properties. Silicon is the second most abundant element on earth. Silicon can be found in sand and rock. In fact, 28 % of the earths crust consists of silicon [10].

Silicones are a class of material that can vary in form from fluids and greases to resins and rubber. Silicone rubber can be purchased in pourable form or as a thick paste. Structurally, silicone relies on a form that is never found in nature. This structure relies on a polymer back-bone of silicon and oxygen atoms arranged in the same fashion as sand and rock. However, the silicon atoms are also joined to organic groups arranged similarly to hydrocarbons found in petroleum and natural gas. Therefore, silicone exhibits the high temperature resistance and toughness known in the mineral world and flexibility and lubricity known in the organic world [10].

Silicones can be categorized into three classifications; polydimethylsiloxanes (MQ or VMQ), polydimethylsiloxanes with phenyl substitutes (PMQ or PVMQ), and polydimethylsiloxanes with 1,1,1 trifluoropropyl substitutes (FVMQ, better known as fluorosilicone) [11]. Phenyl-containing gums are more flexible at low temperature. Vinyl-containing gums offer better compression set resistance.

Organic polymers mainly have an unsaturated double bond in their backbone structure that is prone to attack from oxidation and ozone. The unique chemistry of silicone does not contain an unsaturation in its silicon-oxygen backbone and offers extreme resistance to these factors [12].

Silicone rubber compounds made from gums use inorganic fillers which are not affected by heat. Vulcanization catalysts are usually organic peroxides. Fluorosilicones are used in applications requiring resistance to fuels and better resistance to oils, but at a large cost penalty and a slightly worse high temperature resistance.

Liquid silicone rubber (LSR, or sometimes referred to as LIM-Liquid Injection Molding) is a pumpable silicone having very low viscosity. LSRs usually incorporate a platinum cure system which involves hydrosilylation with platinum catalysts. In addition to an extremely quick cure time, platinum cure systems offer exceptional toughness and tensile strength, but generally require a post cure. Platinum cures are sensitive to cure inhibition from contamination with trace quantities of certain chemicals such as sulfur, or amines commonly used in organic compounds [13].

Silicone applications include: automotive seals and gaskets, spark plug boots, electrical insulators, o-rings, medical seals and gaskets, dampeners, food related parts and similar applications.

3.11   Fluoroelastomer (FKM) [14]

These materials have excellent high temperature resistance and excellent fluid and oil resistance. They are very expensive, but often are the only choice in harsh environments.

There are numerous types of fluoroelastomers, including the following:

- Tetra-fluoroethylene/propylene
- Vinylidene fluoride and hexafluoropropylene copoymers, which were among the first stable fluoroelastomers
- Terpolymers of vinylidene fluoride, tetrafluoroethylene, and perfluoro (methyl vinyl)ether, or PMVE
- Terpolymers of vinylidene fluoride, hexafluoropropylene, and tetrafluoroethylene.

The primary suppliers for FKM in North America are DuPont Dow Elastomers L.L.C. (Viton); Dyneon (Fluorel and Aflas); and Ausimont USA (Tecnoflon).

FKMs have three cure systems available, and some material suppliers include these systems with the compound. Diamine cures were the original cure systems available and are best suited in applications where compression set is not critical and moderate resistance to steam and acid required.

Bisphinol cures offer improved scorch and the best compression set properties. Bisphinol also offers improved steam and acid resistance.

Peroxide cures were originally introduced to improve steam and acid resistance. However, peroxide cures do not offer the compression set resistance of bisphenol.

Applications for FKM include: automotive shaft seals and valve stem seals, fuel related seals, oil field related devices, aerospace seals and gaskets, and related applications.

3.12   Polyurethane (AU and EU) [15]

There are two types of polyurethane rubber used today: polyester-based (AU) and poly-ether-based (EU) products. Polyurethanes posses the best tear strength of any rubber and have excellent wear and abrasion resistance. Polyurethanes have a low threshold of heat toler-ance, and therefore their applications are limited.

Millable PU is cured via peroxides, sulfur, or diisocyanates. There are also castable grades of PU that are poured or injected into molds. These products are often sold as prepolymers that consist of a polyol backbone reacted with diisocyanate.

Polyurethane applications include: roller skate wheels, casters, flex joints, dust boots, grom-mets, medical injection sites, and similar applications.

3.13   Epichlorohydrin (CO and ECO)

Epichlorohydrin has excellent resistance to fuels and is less expensive than fluoroelastomers. It has therefore gained popularity in automotive fuel related applications.

Epichlorohydrin exhibits poor low temperature flexibility, low temperature properties, and abrasion resistance. Applications include: fuel seals, intake manifold boots, and hose.

3.14   Conclusion

A variety of TSE materials exist which exhibit dramatic differences in physical properties, and may require unique processing. Custom molders should understand these differences, and more importantly, they should know how to tweak efficiencies by marrying the material to the molding process.

References

1. Garvey, B. S. Jr., History and Summary of Rubber Technology, *Introduction to Rubber Technology*, Van Nostrand Reinhold Co., NY (1969) p. 2–3.

2. Keenan, R. H., Synthetic Polyisoprene, *The Vanderbilt Handbook* (1978), p. 42–43.

3. Bauer, R. G., Styrene-Butadiene Rubbers, *The Vanderbilt Handbook* (1978), p. 51–52.

4. International Institute of Synthetic Rubber Producers, *Polybutadiene*, www.iisrp.ccm.

5. International Institute of Synthetic Rubber Producers, *Butyl Rubber*, www.iisrp.com.

6. Samuels, M. E., Ethylene-Propylene Rubbers, *The Vanderbilt Handbook* (1978) , p. 47.

7. Dick, J. S., *Rubber Technology: compounding and testing for performance,* Carl Hanser Verlag, Munich (2001), p. 131–132.

8. Dick, J. S., *Rubber Technology: compounding and testing for performance,* Carl Hanser Verlag, Munich (2001), p. 137.

9. Hagman, J. F., Ethylene/Acrylic Elastomer, *The Vanderbilt Handbook* (1978), p. 262.

10. General Electric Co. *GE World of Silicones,* sales brochure.

11. Halladay, J. R., Silicone Elastomers, *Rubber Technology: compounding and testing for performance*, Carl Hanser Verlag, Munich (2001), p. 235.

12. Toub, M., Silicone Elastomers, *Basic Elastomer Technology*, ACS (2001), p. 498–514.

13. Halladay, J. R., Silicone Elastomers, *Rubber Technology: compounding and testing for performance*, Carl Hanser Verlag, Munich (2001), p. 237.

14. Dick, J. S., *Rubber Technology: compounding and testing for performance*, Carl Hanser Verlag, Munich (2001), p. 135–136.

15. Dick, J. S., *Rubber Technology: compounding and testing for performance*, Carl Hanser Verlag, Munich (2001), p. 137.

4 Product Design

4.1 Introduction

It is advisable for product engineers to have experience in the manufacturing sector so that they can design for manufacturability. Designers of TSE articles are confronted with a product that has a difficult to understand manufacturing process. All too often products are over-designed, with too many restrictions that may not be necessary, because the product engineer did not have sufficient training in TSE materials and processes. Over-design can force a higher price versus a product that was designed to simply meet the demands of the application.

Conversely, if an article is under-designed because of the product engineer's inability to understand material and manufacturing limitations, the product can be a complete failure. Therefore, to avoid downstream pitfalls with TSEs, the product designer should use a team approach in designing a given part. The product engineer does not necessarily need to know all of the answers; he just needs to know where to find them. A team, at a minimum, should include the product designer, chemist, tool engineer, process/manufacturing engineer, and quality engineer. Together this team should incorporate the deliverables and checklists outlined in Chapter 14.

This chapter covers basics to assist the product designer of TSE products to design for manu-facturability. The materials and design guides should be used as a reference: the variety of applications for TSEs are far-reaching and cannot be covered in aggregate.

4.2 Material

The product design engineer tasked with designing a custom molded TSE part must start with determining what family of TSEs will satisfactorily operate in the intended environment. This can be as simple as emulating an existing product's material specification that operates successfully in a similar application. Copying or slightly modifying an existing material is the extent of most material explorations. However, if developing a material from scratch is desired, charts as shown in Table 4.1 are readily available and offer a general guideline for potential hosts.

Once a rough list of candidates is determined, the design engineer needs to work in tandem with a TSE chemist to develop a material physical properties specification. Between the chemist and the engineer, they need to decide what physical properties are important, how they translate to the performance of the end product and what maximum/minimum condi-tions are acceptable. The physical properties specification does not guarantee that a material meeting these conditions will work satisfactorily in a given environment. After all, tests used to derive the specifications are performed in a lab environment and molded from slabs and buttons (see Chapter 5). The following ASTM material callout shows an example of a physical properties specification. Unfortunately, unlike the plastics industry where standard materials are readily available along with an array of accompanying physical property test data, TSEs (other than LSR) are all custom mixed.

Table 4.1 Property comparison

Property/TSE	Natural rubber	Polyisoprene rubber (IR)	Styrene butadiene rubber (SBR)	Butadiene rubber (BR)	Nitrile rubber (NBR)	Ethlyene-propylene diene monomer (EPDM)	Butyl rubber (isobutylene isoprene)	Chloroprene rubber (CR)	Chlorosulfonated polyethylene rubber (CSM)	Polyurethane rubber (PU)	Epichlorhydrin rubber (ECO)	Fluorocarbon rubber (FKM)	Fluoro-silicone rubber (FVMQ)	Silicone rubber (Q)	Polyacrylate rubber (ACM)
Weather res.	P	P	P	P	U	E	E	G	G	G	G	E	E	E	E
Ozone resistanc	U	U	U	U	U	E	G	G	E	G	E	E	E	E	G
Heat resistance	U	U	P	P	P	E	G	G	G	P	G	E	E	E	G
Low temp res.	G	G	P	G	P	E	G	G	S	G	G	G	E	E	U
Acid resistance	S	S	S	S	G	E	E	G	G	U	P	G	E	E	P
Alkali resistance	S	P	P	S	G	E	E	G	E	U	S	S	E	G	U
Res. to min. oil	U	U	U	U	G	U	U	G	S	G	G	E	E	U	G
Res. to veg. oil	P	P	P	P	E	G	G	G	G	G	E	E	E	P	E
Abrasion res.	E	E	E	E	E	G	P	G	G	E	P	S	P	U	P
Tear resistance	E	G	P	P	G	P	G	G	P	E	P	P	P	G	U
Strength prop.	E	G	S	G	G	G	S	G	S	E	P	S	U	P	P
Compression se	E	G	G	G	G	G	P	G	S	E	G	G	G	S	S
Impact resilience	E	E	E	G	G	G	G	G	G	E	G	P	P	P	U
Gas barrier proj	P	P	P	P	P	P	E	G	G	G	E	G	U	U	P
Flame resistance	U	U	U	U	P	U	P	G	E	P	G	E	E	P	U
Electrical prop.	G	S	P	P	U	E	E	U	S	G	G	G	E	E	P
Color stability	G	G	G	G	S	E	G	U	E	P	G	P	G	E	U
Tackiness	E	G	P	P	G	P	P	G	E	E	G	G	E	G	G
Processability	E	E	G	P	G	G	U	P	S	U	S	P	G	E	P
Water res.	E	E	G	E	G	E	E	G	G	U	G	E	E	G	U

E = Excellent, G = Good, S = Satisfactory, P = Poor, U = Unsatisfactory

4.2.1   ASTM Classification System for Elastomeric Materials

ASTM D-2000 Specification has been used by the rubber industry to standardize the requirements for physical and chemical properties of elastomeric materials. The specification is referred to as "line call-out". The following "line call-out" is an example of the ASTM D-2000 specification:

2FC 410 A19 B37 G11 Z

The basic requirements of the material is defined by "2FC 410". It can be interpreted as:

2 = Grade number which defines the performance level of requirements

F = The material type. This defines the test temperature requirements. F type material is tested at 200 °C (392 °F)

C = This defines the material class for oil swell requirements (120 % maximum in ASTM # 3 oil)

4 = The target hardness, 40 ± 5

10 = The minimum tensile strength, 1000 PSI

The letters and numbers following the basic requirements are called the suffix requirements:

A = Defines maximum changes in tensile, elongation and hardness of the material after heat aging.

1 = The test method for heat resistance (ASTM D-573)

9 = The test temperature for heat resistance (200 °C)

B37 = Defines the compression set test method and requirement

G11 = Defines tear resistance test method and requirement

Z = Defines special requirements and test method

Large custom TSE mixing companies can be very helpful in offering suggestions and will custom formulate to meet the intended physical specification. Depending on the test requirements, material development can take days, sometimes even months. The chemist needs to pay careful attention when developing the material to match the intended molding method and not just to meet physical properties. This is where the chemist needs to seek the advice from the tool and process engineers. The chemist will draw on his experience to determine what combination of ingredients will satisfy the specification. Several iterations of materials should be used with varying degrees of certain critical ingredients to establish a range. At some point, either the specification is met, or the specification is changed to meet the test results. This may seem to be a compromise, but sometimes, when developing a material from scratch, some expectations may be unachievable. The successful candidate is then evaluated for production mixing compatibility. In some cases, a production batch should be run to verify that all ingredients blend well and that physical properties are met. In addition, the material needs to be monitored throughout the prototype process to confirm physical properties of the end product and processability in molding.

4.3   Design

To understand how to design a custom molded TSE part, the engineer must know about the molding process. As explained throughout this book, TSE has some very unique features — namely elasticity — that are very desirable in the design of seals, gaskets, vibration dampeners, expansion joints, tires, bladders, etc. The nature of TSE's elasticity stimulates some unique and creative designs that cannot be molded in other materials. Undercuts, ribs, and lips (see Fig. 4.1) are squeezed, flexed and distorted during removal from the mold which would destroy the same part molded with other materials. Therefore, the design engineer needs to have an understanding of the chosen TSE's material limitations and how they relate to the mold design, and the intended molding process. For the above mentioned reason, the product designer should consult with manufacturing, process and tooling engineers for guidance. For difficult designs it is often agreed that an educated-guess approach is used and success is determined during prototyping.

Parting line flash, sprue locations, and gates are all irregularities, but necessary to produce a molded part. The design engineer must determine where these irregularities are allowed, and where they need to be avoided. For instance, shaft seals or gaskets should not have any of these irregularities on the seal contact point. Maximum flash extension, sprue/gate size, and protrusion should also be specified.

Figure 4.3 shows a molded part that has four sprues on a rounded top surface. Placing a glass plate over the top surface reveals a dark contrast where contact is made between the plate and the TSE seal. Figure 4.4 shows a close-up view of one of the sprues. Note the leak path formed by the stress concentration at the sprue. A high sprue depresses the surrounding area, inhibiting an effective seal. For the top surface to effectively seal, sprues should be moved to an alternate location, or an alternate fill method should be used.

References to vertical and horizontal flash are common in the TSE industry. Since the TSE molding industry is seated in compression and manual intensive molding, presses were designed vertically. To this day, vertical platforms substantially out-number horizontal press platforms. On the other hand, in the plastic molding industry, vertical platform presses are

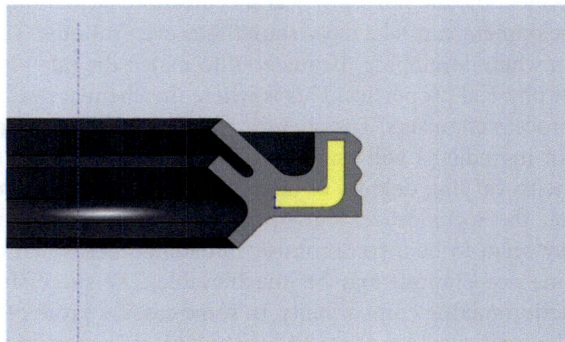

Figure 4.1 Undercuts and ribs

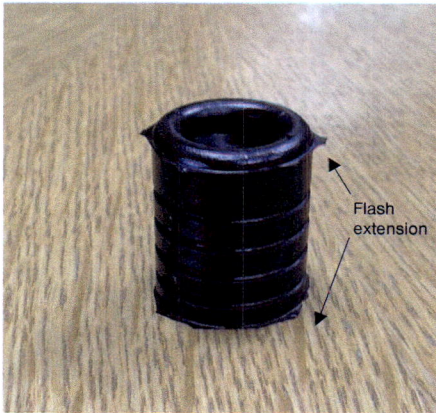

Figure 4.2 Part with horizontal flash extension

Figure 4.3 Glass plate shows leak at sprue

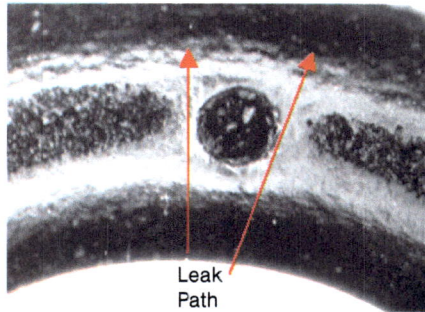

Figure 4.4 Detail of leak path

the exception and are relegated to occasional insert molding. For TSE molding, horizontal flash refers to a mold parting line that is parallel to the press platens on a vertical press. The converse is true for vertical flash. Horizontal flash is preferred and vertical flash should be avoided where possible. Horizontal flash is more controllable and the tool construction has less tendency for damage. The product designer needs to understand flash conditions and other subtleties when designing the product.

Inherent in its elastic properties, TSE is a good candidate for gasket and O-ring sealing applications. Good compression set resistance (described in Chapter 5) allows for constant contact between two flanges and offers sealability. It is important to understand that TSE has limitations when it comes to compression. Generally speaking, a solid cross-section should not be compressed more than 30 % of its original thickness, as described in Fig. 4.5. Compression greater than 30 % can result in splitting of the material, overcoming its tensile strength. Some

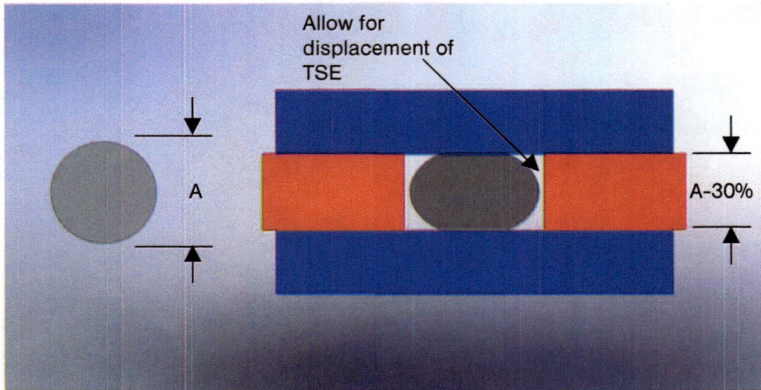

Figure 4.5 Maximum TSE deflection diagram

materials can withstand more compression than others. Consult material suppliers for further detail.

Also, when designing a compression seal (gasket) it is important to understand that space must be provided for the displaced material. The displaced space must accommodate not only the TSE at ambient, but at whatever environmental influences may affect the TSE. For instance, if an O-ring is to be used in a hot oil environment, the displaced area must take into account the thermal expansion of the TSE at the elevated temperature (TSE expands more than steel, aluminum, etc.), and the percentage of swell in the given oil. If the TSE has no place to expand, it will split.

4.3.1   Tolerances

Establishing size limitations for elastomeric parts can be more difficult than for their rigid counterparts. For example, a plastic washer with a 0.240″ inside diameter will not fit onto a 0.250″ steel shaft. The washer needs a diameter callout with a tolerance range that allows it to remain above 0.250″. However, an elastomeric washer with the same dimensions will fit easily onto the shaft by stretching the washer. Does this mean that elastomeric parts can afford to have more liberal tolerances than rigid parts? The simple answer is that it depends. The product design engineer needs to understand what features of the product design are indeed significant, and to what extent.

The tolerance charts in Tables 4.2 to 4.5 are RMA's (Rubber Manufacturers Association's) recommended tolerances for various conditions. The charts are broken down into degrees of precision, A1 to A4. The precision grades are subjective, so to use this chart in determining which category a part may fit into, a manufacturing technique should already be established. For instance, a category A1 part may be molded using a single (or up to four) cavity mold with precision self-registering inserts by either transfer or injection molding. Conversely, a category A4 part may be molded using a 360 cavity cut-in-plate compression mold.

Tables 4.2 to 4.5 RMA tolerance chart [1]. *Courtesy of RMA*

Table 4.2:

STANDARD DIMENSIONAL TOLERANCE TABLE-MOLDED RUBBER PRODUCTS
DRAWING DESIGNATION "A1" HIGH PRECISION [1]

Size (millimeters)		Fixed	Closure	Size (inches)		Fixed	Closure
Above	Incl.			Above	Incl.		
0 - 10		+/-.10	+/- .13	0 - .40		+/- .004	+/- .005
10 - 16		0.13	0.16	.40 - 63		0.005	0.006
16 - 25		0.16	0.20	.63 - 1.00		0.006	0.008
25 - 40		0.20	0.25	1.00 - 1.60		0.008	0.010
40 - 63		0.25	0.32	1.60 - 2.50		0.010	0.013
63 - 100		0.32	0.40	2.50 - 4.00		0.013	0.016
100 - 160		0.40	0.50	4.00 - 6.30		0.016	0.020

Table 4.3:

STANDARD DIMENSIONAL TOLERANCE TABLE-MOLDED RUBBER PRODUCTS
DRAWING DESIGNATION "A2" PRECISION [1]

Size (millimeters)		Fixed	Closure	Size (inches)		Fixed	Closure
Above	Incl.			Above	Incl.		
0 - 10		+/-.16	+/- .20	0 - .40		+/- .006	+/- .008
10 - 16		0.20	0.25	.40 - 63		0.008	0.010
16 - 25		0.25	0.32	.63 - 1.00		0.010	0.013
25 - 40		0.32	0.40	1.00 - 1.60		0.013	0.016
40 - 63		0.40	0.50	1.60 - 2.50		0.016	0.020
63 - 100		0.50	0.63	2.50 - 4.00		0.020	0.025
100 - 160		0.63	0.80	4.00 - 6.30		0.025	0.032
160 - & over				6.30 &over			
multiply by		0.004	0.005	multiply by		0.004	0.005

Table 4.4:

STANDARD DIMENSIONAL TOLERANCE TABLE-MOLDED RUBBER PRODUCTS
DRAWING DESIGNATION "A3" COMMERCIAL [1]

Size (millimeters)		Fixed	Closure	Size (inches)		Fixed	Closure
Above	Incl.			Above	Incl.		
0 - 10		+/- .20	+/- .32	0 - .40		+/- .008	+/- .013
10 - 16		0.25	0.40	.40 - 63		0.010	0.016
16 - 25		0.32	0.50	.63 - 1.00		0.013	0.020
25 - 40		0.40	0.63	1.00 - 1.60		0.016	0.025
40 - 63		0.50	0.80	1.60 - 2.50		0.020	0.032
63 - 100		0.63	1.00	2.50 - 4.00		0.025	0.040
100 - 160		0.80	1.25	4.00 - 6.30		0.032	0.050
160 - & over				6.30 &over			
multiply by		0.005	0.008	multiply by		0.005	0.008

Table 4.5:

STANDARD DIMENSIONAL TOLERANCE TABLE-MOLDED RUBBER PRODUCTS
DRAWING DESIGNATION "A4" BASIC [1]

Size (millimeters)		Fixed	Closure	Size (inches)		Fixed	Closure
Above	Incl.			Above	Incl.		
0 - 10		+/- .32	+/- .80	0 - .40		+/- .013	+/- .032
10 - 16		0.40	0.90	.40 - 63		0.016	0.036
16 - 25		0.50	1.00	.63 - 1.00		0.020	0.040
25 - 40		0.63	1.12	1.00 - 1.60		0.025	0.045
40 - 63		0.80	1.25	1.60 - 2.50		0.032	0.050
63 - 100		1.00	1.40	2.50 - 4.00		0.040	0.056
100 - 160		1.25	1.60	4.00 - 6.30		0.050	0.063
160 - & over				6.30 &over			
multiply by		0.008	0.010	multiply by		0.008	0.010

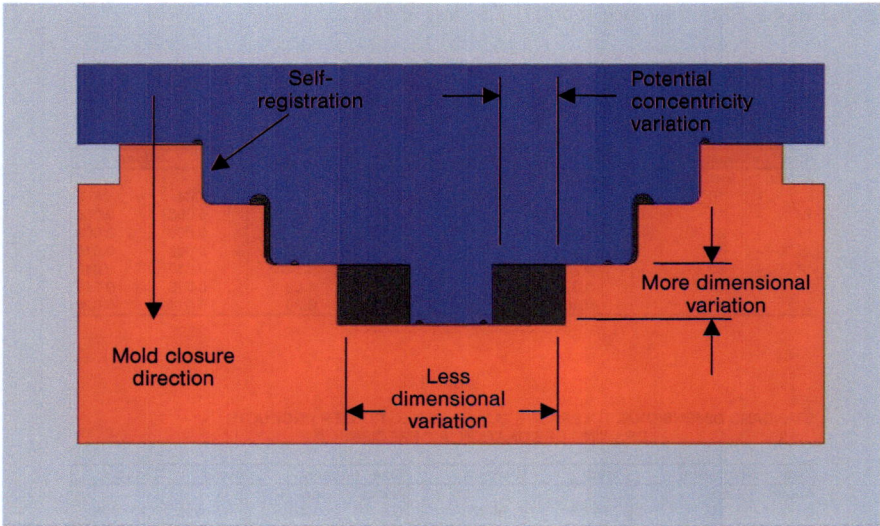

Figure 4.6 Mold concentricity

The RMA charts also draw a distinction between fixed and closure tolerances. Figure 4.6 illustrate a few examples of fixed and closure dimensions as viewed with a part still in the mold. In short, closure dimensions will be parallel to the press closure direction and fixed dimensions will be perpendicular to the press closure direction.

However, dimensions that are created between the two halves of the mold will have greater variation than if the dimension is simply a function of one half of the mold. The RMA charts are intended as a general rule-of-thumb. The product designer needs to consult with manufacturing to fine-tune tolerances to the intended manufacturing process.

The compression mold in Fig. 4.6 illustrates the following potential part measurement concerns. Part measurements taken in the direction of the mold closure will vary to higher degree than measurements that are perpendicular to the mold closure direction. Compression molding will exhibit the most variation in mold closure height given the propensity of flash at the parting line. Concentricity will vary between features molded on opposite mold halves. In multi-cavity molds, each mold insert can be designed to have self-registration to avoid concentricity issues. Molded product variations can be mitigated, but may require more expensive tooling, slower cycle times, more development time, and more scrap. The design engineer must be cognizant of target product cost and timing and recognize where part tolerances are truly critical.

4.3.2   Material Shrinkage

Material shrinkage can be unpredictable. A material's shrinkage factor is derived from a shrinkage mold which has a 10″ long pocket. A strip of uncured TSE is placed in the mold and compression molded. The resultant length of the molded strip is compared to the mold pocket

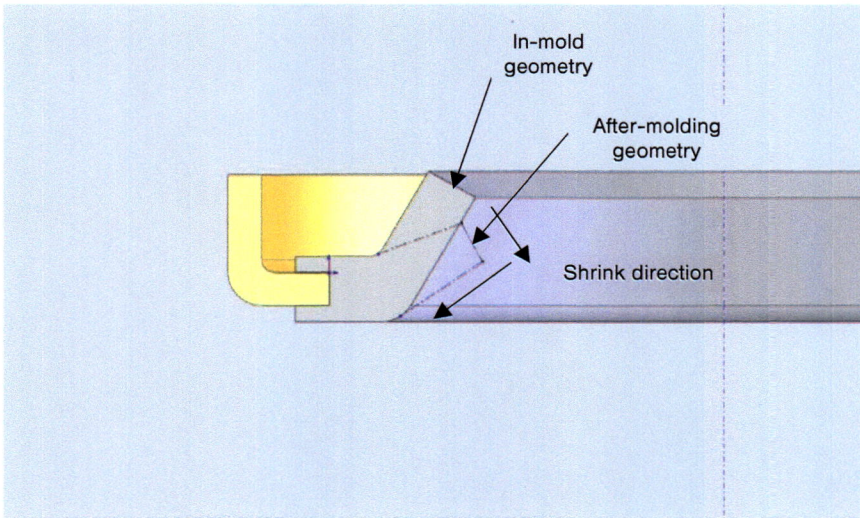

Figure 4.7 TSE shrinkage example

length and a shrinkage percentage is calculated. However, induced material flow stresses can influence shrinkage differently than simply compression molding a strip of TSE. Therefore, shrinkage values should only be used as a guide. Furthermore, bonded carriers can pose challenges to material shrinkage (see Fig. 4.7). The material tends to shrink toward the carrier. Prototyping can often result with several iterations to dial in anticipated shrinkage.

4.4   Conclusion

Product design of TSE's should be undertaken by a designer with experience in manufacturing. In addition, a team effort should be embraced to design a part that can be manufactured cost competitively and operate properly in its intended environment. Product engineers, chemists, manufacturing personnel, and quality assurance should work together to achieve that end.

References

1. The RMA Handbook for Molded, Extruded, Lathecut, and Cellular Products, 6th ed. (2005), Washington D.C., www.rma.org.

5 Material Testing for TSE

5.1 Introduction

TSE is unlike most other materials and requires different test methods and/or alternate interpretations of test data. TSE exhibits visco-elastic properties and can be substantially deformed with relatively low force, while once the force is removed it can have an almost 100 % recovery to its original shape.

There are two reasons for performing material tests: developmental and quality control. Testing is done in a laboratory using uncured materials that are either cured into a form or shape and then physically tested, or placed into an apparatus that records the material's physical properties as it cures.

Developing a material to meet certain physical properties can be a challenging and time-consuming task. Not only does the material need to meet physical and chemical requirements at ambient environments, it needs to be tested in the end product's environment and subjected to its intended abuse. For instance, an O-ring designed to seal gasoline would need to be tested for compression set in a gasoline bath for an extended period of time. Additional complications arise for the chemist when he attempts to match the material's environmental properties with manufacturing processability — a critical step that is all too often overlooked. Viscosity, cure rate, mold fouling, scorch protection, etc., all need to be considered on the processing side of material design.

Quality control testing is done to assure that the compound is properly mixed. Samples are taken at the final stages of mixing and brought to the chemical lab. The representative batch is placed on hold until the test results are proven favorable. At a minimum, sample material is molded into slabs, die cut into dumbbells and tested for durometer, tensile, and elongation. Additionally, uncured material is placed into a rheometer and tested for cure profile. Test results need to statistically fall within control limits to approve the batch. It should be noted that specific procedures exist to gather stock mixing (in-process) samples for testing. ASTM D 3100 to 3109 cover how to gather samples, how long to wait before testing and after curing specimens, etc. For best repeatable results, materials should be handled and allowed to settle in the same fashion every time.

5.2 Physical and Chemical Properties Tests

It is not the intent of this book to detail the various tests available for testing TSE, but to offer a brief description of the most popular tests. Many molding companies that specialize in a certain industry may test to unique industry standards that are uncommon to others. Also, testing of end products is commonly done to confirm the material's ability to perform in the actual application. These tests require specialized equipment and standards and will not be covered.

Common industry standards include:

ASTM   SAE
ISO   JASO
FDA   NSF
USDA

A few of the more common tests provided by the American Society for Testing and Materials (ASTM) are described below. For more detail on any of these tests, consult ASTM.

5.2.1   Tensile Testing

For preparation of this test, the TSE specimen is molded into 6″ × 6″ slabs, 0.080″ thick. Dumbbell shapes are die-cut (see die in Fig. 5.1) from this slab (covered in detail in ASTM D3182). These dumbbell specimens are placed in a tensile tester and stretched to determine tensile strength, elongation, and modulus. Tensile properties testing is described in ASTM D412.

Tensile strength is determined by recording the force per area at rupture. The value is derived from dividing the force by the area of the specimen, before it is stretched.

Elongation is recorded by stretching the specimen and measuring the distance until rupture, and comparing it to the specimen's original length. The recorded value is given as a percentage of original length.

Modulus is simply a value recorded in force per square area to stretch the specimen a predetermined elongation.

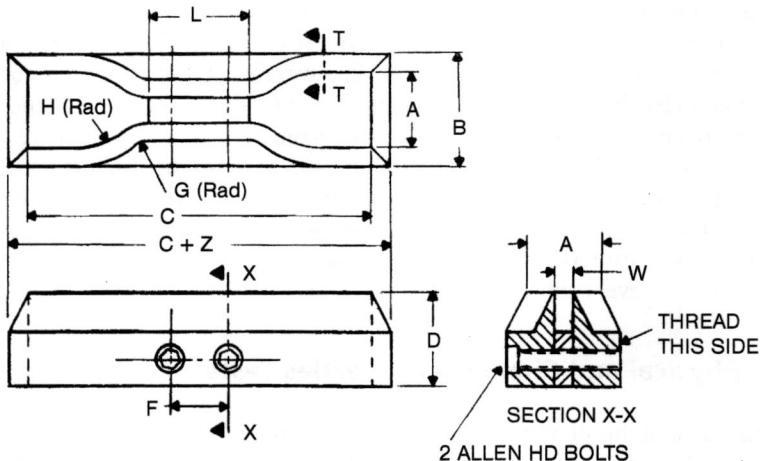

Figure 5.1 Dumbbell cutting die for tensile testing (Courtesy of ASTM International [1])

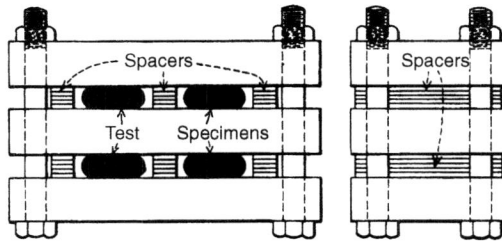

Figure 5.2 Compression set test fixture under constant deflection, Method B (Courtesy of ASTM International [1])

5.2.2   Compression Set

ASTM D 395 describes a test to measure, as a percent, the material's inability to recover over time from a given load (method A) or deflection (method B). Buttons are molded from the test material and compressed over a period of time. 100 % compression set would indicate the button had no recovery; 0 % compression set would indicate that the button displayed full recovery.

5.2.3   Durometer

Hardness, or a resistance to indentation, is designated as a durometer value. ASTM D 1415, and D 2240 describe this test. This test uses a spring loaded stylus which penetrates the TSE surface and records a number on the scale. Several types of durometer testers are available, from hand-held devices to stands with weights attached to act as the indenting force.

Two scales of durometer hardness are used. Shore A is used for soft rubber and shore D is used for hard rubber.

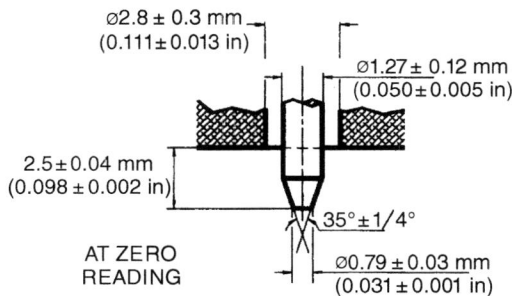

Figure 5.3 Type A and C indenter for durometer (Courtesy of ASTM International [1])

5.3   Heat Aging

For determining physical property degradation by the influence of heat aging over time, test dumbbells are placed in an oven (ASTM D 573) or in test tubes and placed in a block heater (ASTM D 865) at a given temperature for a given period of time. The dumbbells are then allowed to reach room temperature and are tested for physical properties and compared to samples tested without aging.

Other tests can also be performed by heat aging specimens, either fixtured or with slabs, buttons, or dumbbells. For instance, compression set can be tested after aging a specimen in a compression set fixture at elevated temperature for a period of time.

5.3.1   Accelerated Aging

Ideally, material test results would be representative of the intended life of the end product. Unfortunately, the life-span for many products could cover several years. Therefore, accelerated aging tests can be administered. Heat aging is commonly used to accelerate aging, but pressurized air or oxygen are becoming more common. ASTM D 572 and 454 describe these tests. Correlating accelerated aging test results to an intended environment is a real challenge. More often materials are compared to a known satisfactory material — meaning, if a material has performed satisfactorily in a given environment long-term, than that material's original accelerated test results can be used as a benchmark for future qualifications.

5.4   Rubber Property – Vulcanization Using Oscillating Disk Cure

ASTM D 2084 tests for a material's cure profile. An uncured specimen is placed in the cure meter test cavity that is heated to a given temperature. In the center of the cavity is a biconical disk that oscillates to an amplitude of 1° or 3°, and this action exerts shear strain on the test

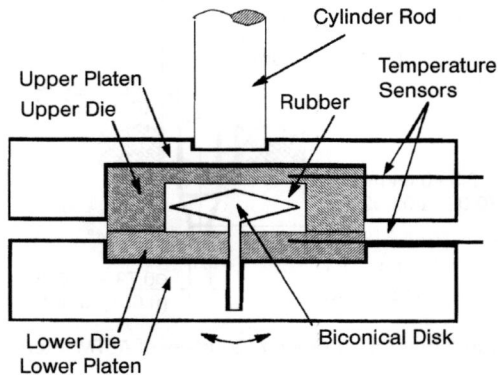

Figure 5.4 Cure meter cavity assembly (Courtesy of ASTM International [1])

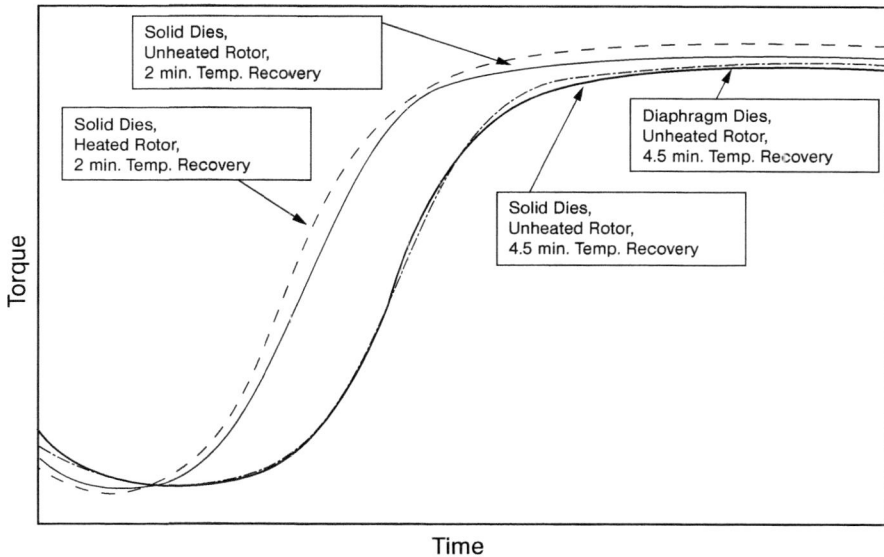

Figure 5.5 Example cure curves from ODR (Courtesy of ASTM International [1])

specimen. This force is continuously recorded as a function of time. As the specimen begins to crosslink, a cure curve establishes its effective cure profile. Maximum and minimum gates can be established to determine if the tested sample will satisfy said controls.

5.5   Fluid Resistance

Most of the above testing can be done with test specimens aged in a fluid bath. ASTM D 471 describes typical fluid aging tests. Often heat and fluids are combined to determine the material's ability to retain it original properties. Fluid swell over time is a common test and predicts the material's inability to resist degradation in certain fluids. It should be noted that fluid aging tests can be dangerous, especially if testing in flammable liquids. Special precautions should be taken. Labs with special safety devices and explosion-proof rooms should be considered for some tests.

Other popular physical properties tests include:

- Ozone resistance ASTM D 1171
- Dynamic flexing ASTM D 430
- Abrasion tests ASTM D 1630
- Tear resistance ASTM D 624

5.6   State-of-Cure

Verifying whether a molded article is completely cured is more difficult than simply relying on the fact that the article appears rigid enough. If articles are to be post-cured, it may be a moot point. However, many TSEs do not require a post-cure. In these cases state-of-cure tests can be performed.

Hardness

Durometer testing, as illustrated in the ASTM tests earlier, is perhaps the easiest method to determine if a molded article is completely cured. Instead of slabs or buttons, actual molded articles are measured for hardness with a durometer tester. Hardness is, first of all, a subjective test. Secondly, it may be difficult to find a location suitable and repeatable to place the stylus. Lastly, thickness of the article has an effect on measurements. Hardness testing can be performed as a quick check method to verify state-of-cure, but it should be noted that it is not an exact test method.

Cross-Link Density

Cross-link density is a test that measures the amount of cross-linking that has taken place in a molded article. This test measures the initial density of the molded article. Then, the article is soaked in a given solvent and allowed to swell. The article is then removed and allowed to rest, and remeasured for density. The difference in density is compared to the raw polymer to determine if adequate crosslinking has taken place.

Free-Sulfur Content

Free-sulfur is the amount of sulfur available in a rubber compound available for further vulcanization that is extractable by sodium sulfate. Measuring free-sulfur determines if the molded article achieved adequate state-of-cure [1].

Compression Set

Molded articles can be placed into a fixture similar to that illustrated in the ASTM test for compression set and tested at elevated temperatures. This is a good and inexpensive test for articles that have a solid and consistent cross-section. Portions of a molded article can be cut and placed into the fixture for testing. Drawbacks to this method include: destructive testing, long test time, and subject to suitable part geometry.

5.7   Conclusion

Testing of TSE materials and finished products covers a vast array of possibilities. The tests and methods described in this chapter are only a small sampling for the custom molder. It is a good idea to order ASTM standard test method volumes and become familiar with them. Custom molders will invariably require some testing from these standards. In addition, special industry test methods should be understood and be included in advanced quality planning.

References

1. Copied with permission from ASTM International, West Conshohocken, PA.

6 Polymer Flow

Flow of TSEs is a complex subject. Since TSEs need to displace and travel during any particular molding process, understanding the fundamentals of TSE flow is vital in establishing a new molding process, or troubleshooting/optimizing an existing one. To add to the complication of flow, TSEs flow changes at an irreversible rate once crosslinking occurs. Several tests are available to characterize flow properties both prior to and during crosslinking. However, much of this science remains cloaked in uncertainty. TSE molders continue to use test data as a guide instead of a rule. Proof of concept is still delivered in the molding operation. This chapter will provide an understanding of TSE flow and how it can be deciphered to optimize a process.

6.1 Viscosity

According to Webster, viscosity is the internal friction of a fluid, caused by molecular attraction, which makes it resist a tendency to flow [1]. So, if a stress is applied to a fluid, viscosity is the material's ability to resist movement. The formula for viscosity (η) is:

$$\eta = \frac{\text{shear stress (applied force)}}{\text{shear rate (rate of deformation)}}$$

Shear stress is defined as a stress applied parallel to a face of material. The formula for shear stress (τ) is:

$$\tau = \frac{\text{shear force at that location}}{\text{area of section parallel to shear force}}$$

Shear rate is defined as the velocity over the given cross section with which molten or fluid layers are gliding along each other. The formula for shear rate (γ) is:

$$\gamma = \frac{\text{velocity}}{\text{thickness}}$$

The standard SI measure for viscosity is Pascal·second (Pa·s). However, Pa·s is not a measure typically embraced by most industries. Poise (P) is a more commonly used measure of viscosity which is dyne·second per square centimeter (dyne·s/cm^2). Ten Poise equals one Pascal·second and therefore a centipoise (cp, or cps) and millipascal·second (mPa·s) are identical.

$$1 \text{ Pa·s} = 10 \text{ P} = 1\,000 \text{ mPa·s}$$

$$1 \text{ cps} = 1 \text{ mPa·s}$$

Viscosity of TSEs, as with most polymers, varies significantly by altering temperature and shear stress, which is what TSEs encounter during any molding operation, and therefore requires a method to measure the material's changes throughout these variables. Traditional Poise viscosity measurements only record viscosity under single conditions. Furthermore, the magnitude of viscosity of Poise for TSEs, with perhaps the exception of LSRs, is impractical for testing. Therefore, alternate test methods were developed specifically for measuring the viscosity of TSEs, and will be discussed later in this chapter.

For purposes of comparison, following are some common materials and their viscosities:

Material	Approximate Viscosity (cps)
Water at 70 °F	1–5
Motor oil SAE 30 or maple syrup	150–200
Molasses	5 000–10 000
Tomato paste or peanut butter	150 000–250 000
Caulking compound [2]	5 000 000–10 000 000
LD polyethylene plastic (300 °F)	10 000 000
Polyamides (500 °F) [3]	10 000 000
Millable organic TSE	20 000 000–100 000 000
Millable silicone TSE	5 000 000–20 000 000
LSR (liquid silicone rubber) [4]	300 000–1 000 000

6.2   Elasticity [5]

A perfectly elastic material will conform to a deformation under force and have 100 % recovery once the force is removed. Elasticity is one of the properties desirable in TSE materials. A truly elastic material (such as some metals at very low strains) will conform to Hooke's Law:

$$\sigma = E\gamma$$

Where:

σ = stress, or force per unit area
γ = strain, or displacement, as measured from change in length
E = the static modulus of elasticity

As mentioned previously, TSE materials are visco-elastic. They do not exhibit perfect elastic characteristics, which have no measured stress response as the rate of applied deformation varies. As TSE materials cure, they transition from a predominant viscous state to an elastic state. Yet even in the uncured state, TSEs will exhibit elasticity due to entanglement of long molecular chains. This elasticity among uncured TSE is called *nerve*. Uncured TSEs with high nerve (high elasticity) have a tendency to resist processing. It is important to understand the nerve of a TSE so that it can compliment the desired mixing and molding processes.

6.3 Plasticity [6]

Plasticity is the opposite of elasticity. A completely plastic body will exhibit no recovery when deformed by shear stress. The properties of a vulcanized TSE tend to approach that of a completely elastic material. There is currently limited commercial use for a TSE exhibiting completely plastic characteristics.

6.4 Rheology

Rheology is the study of change in form and flow of a material in terms of elasticity, viscosity, and plasticity. Rheology is affected by the material's chemical composition, time and temperature dependency, and shear.

There are only two types of fluids: Newtonian and non-Newtonian. Uncured TSEs and molten thermoplastics are considered fluids, and are therefore governed by the Power Laws of fluids.

Newtonian Fluids

These fluids are considered ideal liquids and truly viscous. As shear rate changes, the viscosity remains the same for these fluids. Water, and some oils or solvents display this characteristic, and are therefore Newtonian fluids. A Newtonian fluid is considered one-dimensional in regard to viscosity: temperature is the only factor that will affect a Newtonian fluid's viscosity.

Non-Newtonian Fluids

These fluids are affected by shear. Many non-Newtonian fluids, at a molecular level, can be substantially rearranged in flow. Non-Newtonian fluids are categorized into two Power Law fluids:

Class I: Time-*Independent* Non-Newtonian fluids

- Pseudoplastic: As shear increases, viscosity decreases. TSEs, most thermoplastics, shampoos and paints behave in this manner. These fluids are also called shear-thinning fluids.
- Dilatant: As shear increases, viscosity increases. Examples include corn starch and wet sand. These fluids are also called shear-thickening fluids.
- Plastic (Bingham) fluids: behave like solids until a critical shear rate is achieved where the material starts to flow and then may exhibit Newtonian, dilatant, or pseudoplastic characteristics. Examples include toothpaste, hand cream, chocolate, and grease [7].

Class II: Time-*Dependent* Non-Newtonian Fluids

- Rheopectic: Viscosity increases as shear increases as a function of time, but as shear is decreased, the material gradually recovers to its original properties. True rheopectic fluids are rare. Example include gypsum paste and painter's ink.

- Thixotropic: Viscosity decreases over time given a constant shear rate. As shear decreases, the material gradually recovers to its original properties. Examples include yogurt and paint. Pseudoplastic and thixotropic have similar characteristics and are often, incorrectly, used interchangeably [7].

6.4.1   Thermoplastic Fluid Properties

Most thermoplastic materials in their fluid (molten) state are considered non-Newtonian Class I, pseudoplastic fluids.

As indicated in Fig. 6.1, as shear rate increases, the fluid's viscosity decreases. In thermoplastics, typically as shear increases, the curve becomes more linear, and in rare cases, it approaches Newtonian behavior [8]. In any event it is *not* time-dependent. Most pseudoplastic curves also become more linear with extremely high shear rates. Nevertheless, shear has a greater influence on viscosity than temperature for thermoplastics, as indicated in the two graphs in Fig. 6.2.

The viscosity of thermoplastics is difficult to measure in standard terms such as Poise or Pascal · seconds. Viscosity is most often given in a number derived from a melt flow index machine (see Fig. 6.3). This machine measures the rate of molten material that is extruded through a given diameter and length of an orifice. Temperature, force (weight), and piston position are variables in the timed test. Melt flow is covered in ASTM D1238 and ISO 1133.

Barring any chemical degradation, the viscosity of plastics can fluctuate back and forth as established by the temperature and shear of the melt. Furthermore, residence time (time spent in the injection molding machine or hot runner), again barring any chemical degradation, is important, but not to the critical extent exhibited in thermosets (as will be described

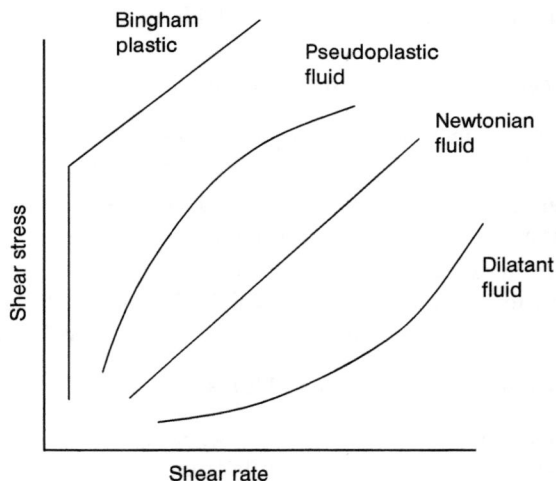

Figure 6.1 Fluid and shear

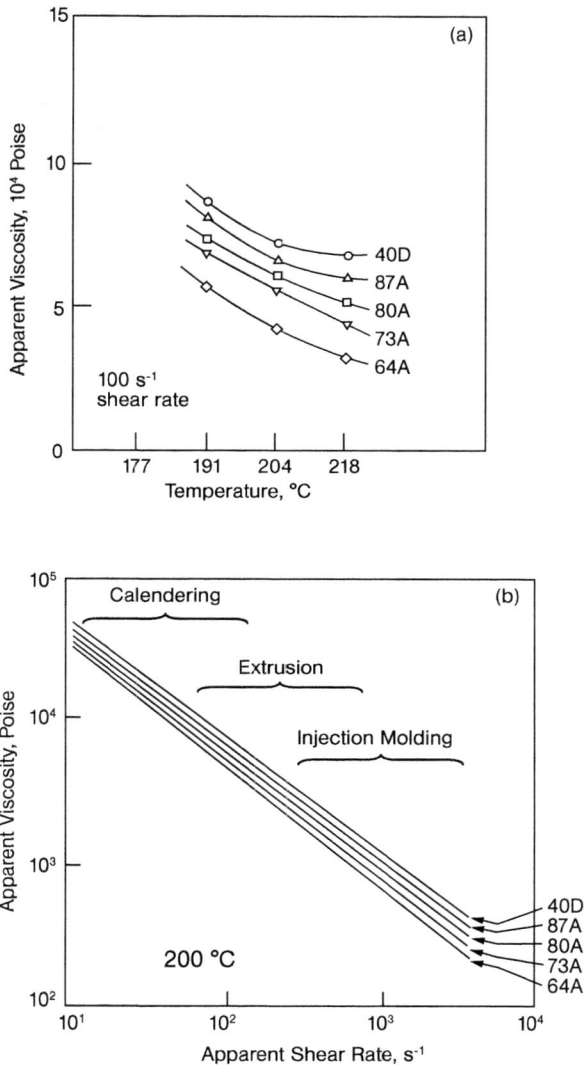

Figure 6.2 Viscosity differences in TPVs with (a) temperature and (b) shear rate [9]

later). Thermoplastic flow can be said to be heavily influenced by shear. Optimum flow of thermoplastic materials is best achieved by shear rather than simply applying heat.

Once the viscosity achieves its predictable state, this plastic shear rate curve can be effectively loaded into a computer-simulated flow program. Computer-simulated flow programs can be a useful aid in designing runner and gate systems for plastic injection molds.

Figure 6.3 Melt flow index machine LFI-3000 (Courtesy of Qualitest international, Inc.) [10]

The following provides an overview of the flow properties as they relate to a plastic injection molding process.

- Plastic pellets, a solid, are transported into the injection throat. Through an opening in the barrel, the pellets are fed down the barrel by an auger.
- The barrel is generally only heated to control the plastic temperate. It is best to generate shear from the auger (plasticizing screw). The resultant shear and heat combination transform the solid into a fluid.
- The plastic is now molten or a fluid and it exhibits the characteristics of a non-time dependent, non-Newtonian, (or in rare cases Newtonian) fluid.
- In this state, it can be pumped through the injection unit, nozzle, and hot runner system with little difference to viscosity, provided the path through which the material flows is similar in temperature and shear heat is held to a minimum.
- The molten plastic reaches the mold cavity. The mold cavity temperature is maintained below the plastic melt point.
- The material solidifies in the mold cavity and is ejected from the mold.

6.4.2   TSE Fluid Properties

From a TSE molder's perspective, rheology is the single most important property that can influence the processing of TSEs. Much is still to be learned about the flow characteristics of TSE. Interestingly, uncured TSE is considered a fluid. An uncured TSE specimen placed on a flat surface will eventually (albeit at a glacial speed) displace and flow (time, pressure and temperature will influence this process). Of all the polymers available, TSEs exhibit the highest molecular weight. High molecular weight translates into higher viscosity [11].

The flow of TSEs is considered non-Newtonian. TSEs initially exhibit pseudoplastic behavior, but over time transitions to a state similar to rheopectic. In other words, as shear is introduced, viscosity decreases; but over time, chemical crosslinking occurs that will increase the viscosity at a non-reversible rate (technically, rheopectic fluids resort back to their original viscosity once shear is removed). The chemical crosslinking that occurs prior to the prescribed timing is considered scorch. This phenomenon is what sets thermosets apart from thermoplastics as very difficult to predict, and as a result difficult to process.

Viscosity of a TSE is predominantly predicated by its molecular weight, but can be altered by modifying levels of fillers, plasticizers, lubricants, or processing oils. The base polymer can usually be purchased in a variety of viscosity ranges. Viscosity of TSEs are much higher than their thermoplastic counterparts, particularly when considering the temperature at which they are required to flow (consider a cold runner for TSE, vs. hot runner for thermoplastic).

6.5   Shear Thinning

Shear thinning is the effect of a decrease in measured viscosity as the shear rate increases, which is pseudoplastic or thixotropic behavior. The rate of viscosity change vs. shear varies with different TSE materials and it is important to draw distinctions to provide the optimum

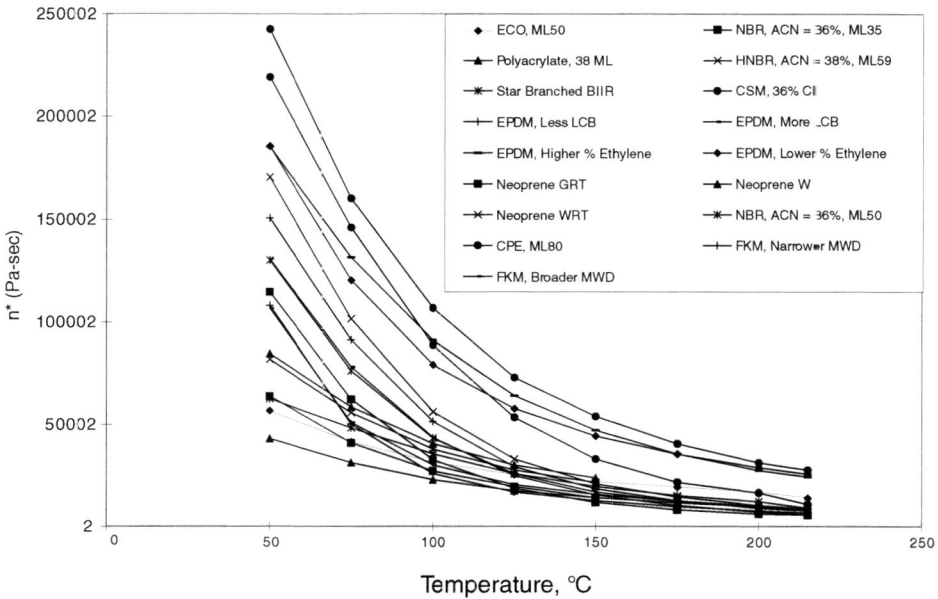

Figure 6.4a Effect of temperature on viscosity for various TPEs [12]

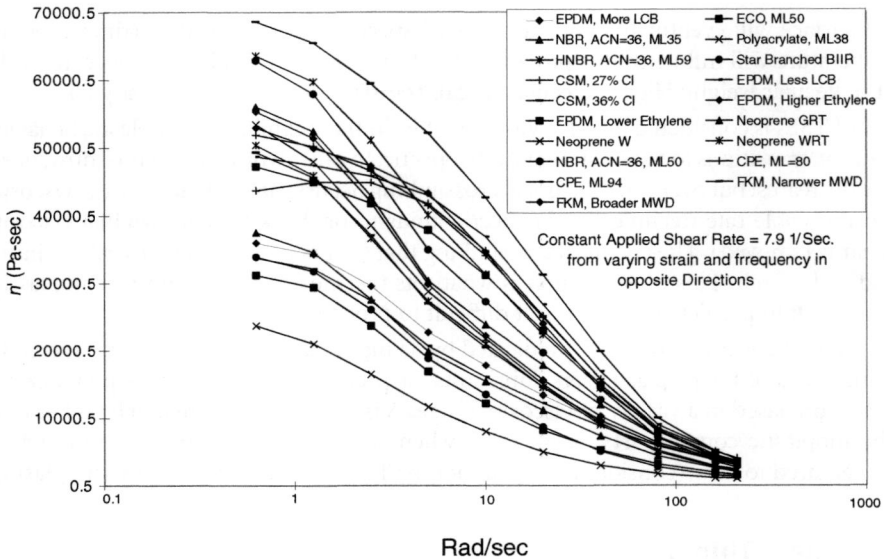

Figure 6.4b Effect of shear on viscosity for various TPEs [12]

manufacturing process given a material's viscosity change. Shear thinning behavior for TSEs is measured by rotational viscometers, capillary rheometers and/or oscillating rheometers. The charts in Fig. 6.4a show viscosity change as it is effected by temperature and shear.

6.6   Rotational Viscometers

Melvin Mooney developed a testing apparatus in the 1930s that set out to measure TSE flow. This machine simply incorporates a rotor with a gear-shaped outer diameter, rotating at a given RPM, that is sandwiched between two mold halves. The cavity in the mold halves also incorporates grooves to prevent the specimen from slipping during the rotor rotation. Two pre-cut specimens with a combined volume of 25 cm^3 are placed in the mold. The mold halves are closed under pressure and the TSE fills the cavity. Usually there is a preheating time of one minute to allow the TSE to reach the set temperature. The rotor is then activated and turns at two revolutions per minute (2 RPM). The test records viscosity in Mooney units (MU), which are arbitrary units based on torque.

Figure 6.5 depicts a Mooney viscosity chart. A typical test result may be displayed as 50 ML (1 + 4). This indicates that a large standard rotor was used (representing the ML) as opposed to a small rotor (MS). The test was run with a one minute preheat (1) and the rotor was turning for four minutes (4). The lowest recorded viscosity during the last 30 seconds of the test was 50.

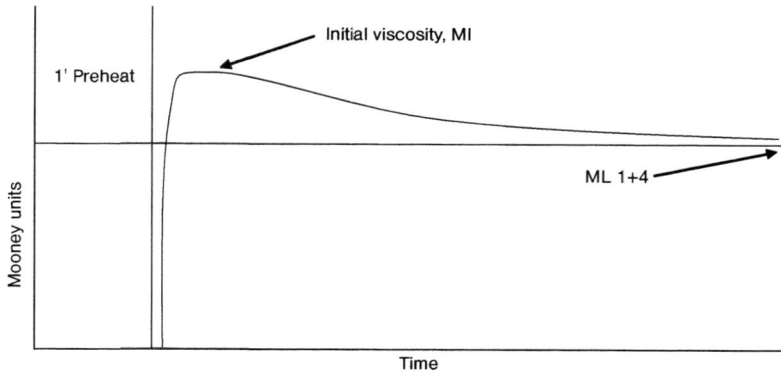

Figure 6.5 Mooney viscosity chart

Mooney viscosity can be measured at various temperatures and can measure with or without crosslinking chemicals in the TSE. Usually temperature is set at the TSEs processing temperature, such as the injection chamber temperature of an injection molding process.

Mooney scorch is performed at a temperature above 100 °C (typically a temperature where the TSE will reach scorch between 10 and 20 minutes). This test is generally performed with the large standard rotor and measures the amount of time required for the TSE to rise above the minimum viscosity by five Mooney units. Figure 6.6 shows a typical Mooney scorch chart.

Mooney viscosity is still used as a guide for viscosity of TSEs. A problem with Mooney results is that the shear rate used during the test (1 s^{-1}) is much lower than what actually occurs in most molding operation [13].

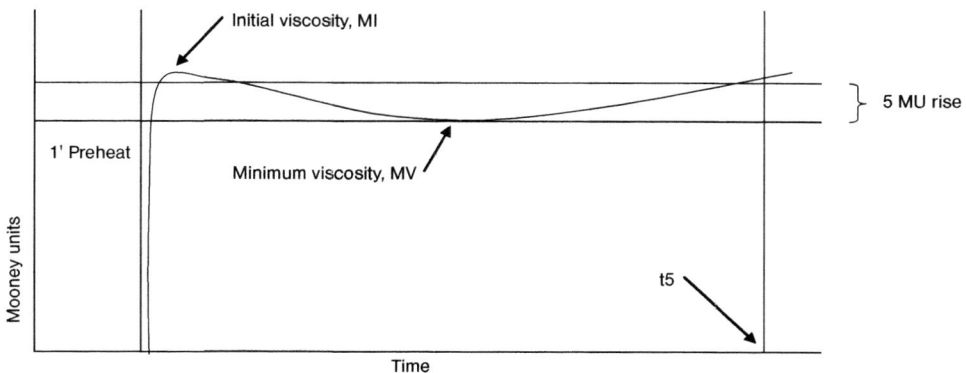

Figure 6.6 Mooney scorch chart

6.7   Oscillating Rheometers

Rheometer testing is used to determine the selected vulcanization characteristics of organic and silicone TSE compounds. The rheometer is an excellent tool for R&D compound development and provides an efficient test method for quality control of production batches of mixed TSE compounds.

There are various oscillating rheometers. The most common are ODR (Oscillating Disk Rheometer) and MDR (Moving Die Rheometer). For discussion purposes, the ODR will be described. The summary of rheometer testing consists of a test specimen of a vulcanizable TSE compound that is inserted into the test cavity and, after a closure action, is contained in a sealed cavity under pressure. The cavity is maintained at an elevated temperature. The TSE completely surrounds a conical disk after the die halves are closed. The disk is oscillated through small rotational amplitudes (1 or 3 degrees) and this action exerts a shear strain on the test specimen. The force required to oscillate or rotate the disk to a maximum amplitude is continuously recorded as a function of time. The force is proportional to the shear modulus (stiffness) of the test sample at the established test temperature. This stiffness initially decreases as it warms up (ML); then it increases to vulcanization (MH). The test is completed when the recorded torque either rises to an equilibrium or maximum value, or when a predetermined time has elapsed. The time required for testing a sample is a function of the characteristics of the TSE compound and of the test temperature [14].

Rheometer testing is a method to analyze the flow of the uncured TSE, but is only an indication of what may happen in a molding environment. Mostly, flow characteristics are compared to a known acceptable cure curve. In other words, if a material displays satisfactory flow/cure

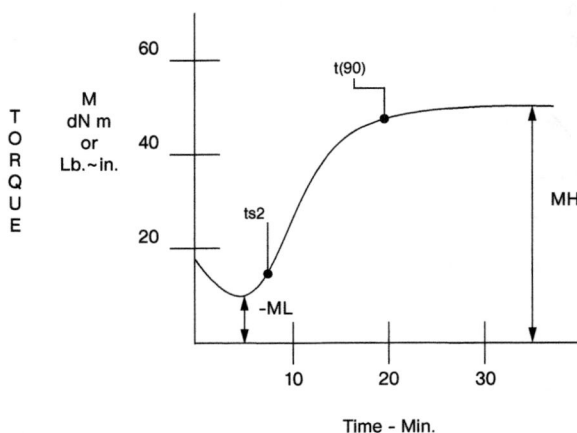

Figure 6.7 Oscillating Disk rheometer (ODR) curve; ODR standardized values:
ML = Minimum torque (lowest viscosity)
MH = Maximum torque
TS1 or TS2 = Minutes to torque rise above ML (measure of scorch)
T90 = Cure completion (more specifically 90 % of cure)

characteristics during a given molding process, its cure curve is established as the standard. The transition and interaction between pseudoplastic and rheopectic-like flow, and the plethora of potential shear related factors involved during a molding process, renders computerized flow simulators ineffective.

The above mentioned intrinsic complications in the flow of TSE foster a hit-or-miss approach to matching a material to a molding process; perpetuating the notion that rubber molding is more of an art than a science. Experience is key to matching the material to the process. Material-to-process matching may reflect on one's past experience in posing the following questions/comments:

- How did this material run in a similar process?
- What can we do to tweak this material to run in this process?
- Let's try it and find out.

The rheometer curve is still the most popular method for fingerprinting the processability of TSE material. There are upper and lower control limits on this curve which determine if the material is acceptable or not. However, molding tests should be performed to establish upper and lower control limits as they relate to the actual molding process.

6.8   Conclusion

Understanding TSE flow may never be completely predictable in a given molding environment, but test equipment continues to improve to offer greater predictive data on shear, temperature, and cure interactions as they relate to flow. RPA (Rubber Process Analyzer from Alpha Technologies) and DMA (Dynamic Mechanical Analyzer by Perkin Elmer) machines have evolved to offer insight into dynamic physical property changes that best simulate TSE environments during a molding process. As the material is the greatest source of variation in a TSE molding operation, emphasis should be placed on statistically determining how a given material will react in a given molding process. The key to the success of a custom molder will be to include a statistically proven material acceptance plan.

References

1. Guralnik, D. B., *Webster's New World Dictionary of the American Language* (2nd ed.). Simon and Shuster. (1982).

2. Retrieved on (9/6/08) from www.research-equipment.com/viscosity.

3. Retrieved on (9/6/08) from www.invibio.com, PEEK-Optima Polymer Processing guide.

4. Toub, M., Silicone Elastomers, *Basic Rubber Technology*, Rubber Division of the American Chemical Society (2001), p. 498–514.

5. Dick, J. S., *Rubber Technology: compounding and testing for performance*, Carl Hanser Verlag, Munich (2001), p. 29–30.

6. Widenor, W., Plastometers and Viscometers, *The Vanderbilt Handbook* (1978), p. 597.

7. Swearingen, C., Velocity-Profile Deviations Influence Flowmeter Performance, *Cole-Parmer technical Library* (2000).

8. Retrieved on (8/1/08) from www.chem.com.au/science/rheology.

9. Dick, J. S., Rubber Technology: compounding and testing for performance, Carl Hanser Verlag, Munich (2001), p. 279.

10. Qualitest North America, Plantation, FL.

11. Putman, M., Richter, P., The use of Rheological testing for predicting processability, *Rubber World* (May, 2007).

12. Dick, J. S., Comparison of shear thinning behavior of different elastomers using capillary and rotorless shear rheometry, Presented at a meeting of the Rubber Division, ACS, Dallas, Texas 4/4/00 to 4/6/00.

13. Retrieved on (8/1/08) from www.chem.com.au/science/rheology, p. 22–23, 34.

14. Sadr, F., Personal interview 4/15/08.

7   Molding Methods and Related Topics

7.1   Introduction

TSE molding is generally a more complex method of molding when compared to plastic injection molding. With in-house mixing capabilities, most TSE molding companies have their own standards when it comes to molds, processes, and processing equipment. Given the proclivity to automation and quick cycle times, conventional wisdom would dictate that injection molding should be embraced for all custom TSE molding. However, for good reason, it is still not the predominant molding method. Custom TSE molders will typically have capabilities of compression, transfer, and injection molding, and often hybrids of each type. Each company also has its preference of molding processes with matching unique material formulations, making it difficult for press manufacturers to develop standards in the industry. Hence the secretive stigma associated with the rubber molding industry.

In contrast, the plastic industry almost exclusively embraces the injection molding process for custom molded parts. In the plastic injection molding industry, mold bases are either purchased or fabricated to specific standard sizes, and are equipped with standard components. These mold bases will fit into most major press manufacturer's equipment as their machines are built to accommodate these standards. Industry standards, guidelines and advice are readily available in the plastic injection molding industry, but are difficult to come by in the TSE molding industry. Lack of standards, a variety of molding methods, and in-house material development all contribute to the unique nature of TSE molding.

This chapter will briefly discuss the advantages and disadvantages of various molding methods. Details of each method will be covered in subsequent chapters. The intent of this chapter is to cover techniques that may be common to various methods. Mold construction, including part ejection/removal, heat transfer, vacuum, and platings/coatings will be discussed. Mold sequences and operations will be covered to provide an understanding of potential efficiency improvements.

7.2   Choosing a Process

Figure 7.1 indicates general categories for molding methods and how they relate in complexity to cure times and material viscosities. In many cases, general purpose gum material compounds can run in compression, transfer, or injection molds. However, compounds are best if formulated to match a desired molding method.

Many parameters need to be evaluated before determining which molding method best suits the part to be molded. End product cost is always a major factor in determining the most economical process. However, what is initially perceived as the most economical process is not always the best choice. For example, would compression molding be an obvious choice for high-volume, high-polymer cost, and high-precision shaft seals? Perhaps surprisingly, most

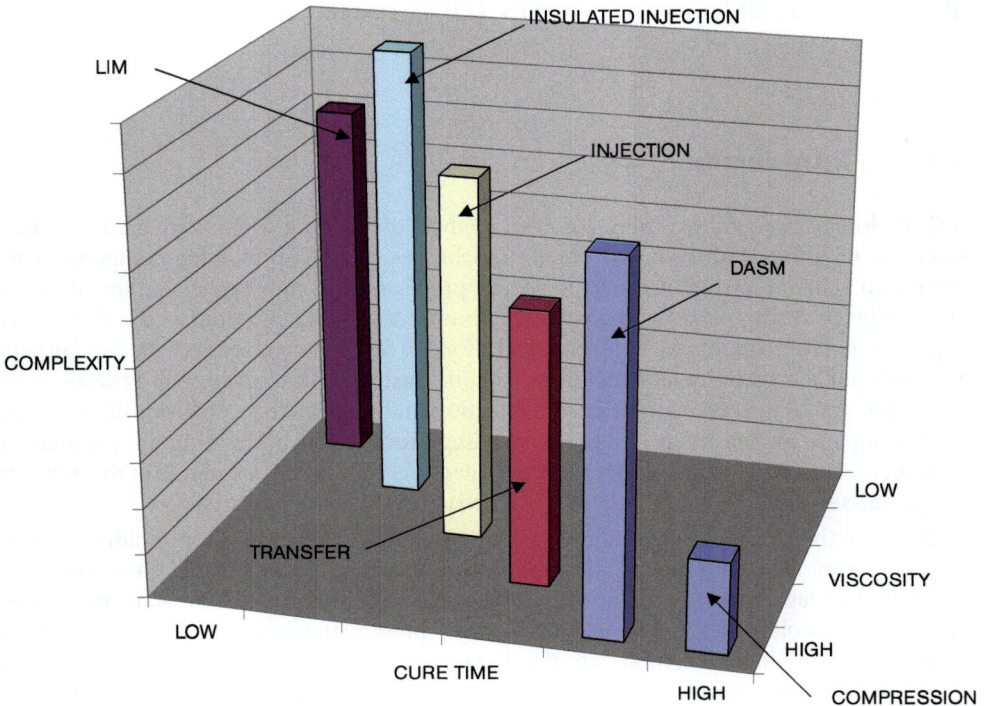

Figure 7.1 Molding type vs. cure time and complexity

of these seals are compression molded and for good reason. Precision donut shaped preps and highly engineered compression tooling has proven to be the best and most economical method for these types of parts. Would transfer molding come to mind for small precision seals? Again, the perception may be that it is an archaic process, but transfer molding offers very high potential cavitation, flash-free parts, and a very forgiving process.

Some molding companies specialize in certain material types. For example, LSR has become a favorite material for the medical molding industry. For such companies, molding choices are limited. However, most TSE molding companies offer a variety of compression, transfer, and injection molding. Table 7.1 is a chart showing product application types and most likely process candidates. This is only a general guide. Many other factors may influence the decision on which process to choose.

As mentioned earlier, automated processes are still uncommon in TSE molding. In non-automated molding, an operator is required to load TSE into the mold (if compression or transfer molding is used), load inserts (if bonded carriers are used), remove flash, manipulate various mold plates, demold parts, brush vents/lands, air blast parting line, and spray lube (if required). Each one of these steps offers potential variation in the molding process.

Table 7.1 Comparison of molding methods

Parameter	Condition	Molding Process Choice						
		compression	dual acting compression	transfer	cold pot tranfer	injection	cold runner injection	valve gate injection
Volume (press utilization)	High	×	×	×	×	×	×	×
	Low	×	×	×	×	×		
Polymer cost	High	×	×		×		×	×
	Low	×	×	×		×		
Part tolerance	High		×	×	×	×	×	×
	Low	×	×	×	×	×	×	×
Flash tolerance	High			×	×	×	×	×
	Low	×	×	×	×	×	×	×
Polymer viscosity	High	×	×	×		×		
	Low	×	×	×	×	×	×	×
Part size	High	×	×			×	×	×
	Low	×		×	×			

Being almost 300 °F above ambient, mold temperatures can fluctuate tremendously during a molding cycle — particularly if thin plates, and/or difficult demolding/cleaning are required.

Successfully molding TSE requires consistency in every venue. This cannot be over-stressed. From the mixing of the TSE to the demolding of the end-product, variation needs to be mitigated. A perfectly tuned/consistent molding operation will be relegated to sub-par, if material variation is present. In the same fashion, spending premiums on perfectly mixed material used for a manually operated molding process could be throwing good money after bad.

Mold temperature is a critical factor in TSE molding. A 10 °F temperature dip in the mold can create an undercure, and a 10 °F rise in mold temperature can cause scorch, or non-fills. Molds that require lengthy demolding times can easily see 20 °F temperature swings throughout the molding cycle.

It is impossible to run a molding operation without variation. Some variation is acceptable — it is a consistent (repeatable) and a predictable variation that the TSE molder seeks. Naturally, automated systems will provide more consistent variation than a manual operation. In a non-automated process, mold-open time (or change time) can fluctuate markedly, despite operator best efforts to consistently open the mold, discharge the parts, clean the mold, and load inserts and/or preps cycle to cycle. Automation eliminates this operator variation.

Every operator may have a slightly different technique, or may have difficulty with an article stuck in the mold, and all may contribute to temperature variation. Even automated molding operations can display unwanted variation. If a mold is clamped and not cycled for a period of time, temperature saturation takes place. Meaning, mold plates will heat up more than if the mold was running and cycling. Therefore, once started, molds should run continuously through breaks and lunches. In the event a mold press needs to be shut down, stabilizing cycles should be administered before production commences.

Degrees of automation should be considered where feasible. Not every part can be molded automatically. However, the process/tooling engineer needs to consider the complications inherent in TSE molding and to mitigate potential influences of variation. This can be accomplished by automating certain aspects of the molding cycle and minimizing mold cavitation. For instance, automated plate handling, K. O., and material handling can all have a significant positive effect on variation.

It stands to reason that injection molding, and particularly insulated material delivery systems, offer greater potential for automation. Material delivery and runner/sprue removal concerns are eliminated/minimized with valve gated injection molding. The engineer needs to consider automation where practical to minimize variation to reduce scrap and strive for the most cost-efficient cycle times. Automation, valve gated injection molds, injection molding machines, and other equipment can require large capital expenses. Obviously, a return on investment needs to justify any equipment expenses.

7.3   Book Mold

A book mold can be used for compression or transfer molding. The mold is designed to be moved, in its entirety, in and out of a vertical press for every cycle. Once the cure cycle is completed, the press opens and the entire mold is shuttled out to a lift table in front of the press; usually alienating itself from the heat source. A lift plate is activated and separates the mold plates. The operator then swings the individual plates away from each other, further influencing heat loss. The center plate (which usually houses the molded articles) is swung to one side and a K. O. plate is activated that forces the articles from the plate (see Fig. 7.2).

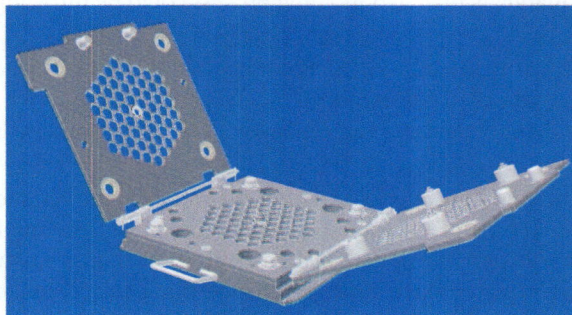

Figure 7.2 Typical book type mold (Courtesy of Ft. Wayne Mold [1])

During the mold-open cycle, molded articles are removed, the mold is brushed, flash is removed with an air blast, and preps and inserts to be bonded (if used) are loaded into the mold. Heating platens are set at a much higher temperature than what is called for in the mold. As the mold is clamped in the cure cycle, it is recovering heat loss, resulting in uneven or shifting mold temperatures during material cure. Long change times harbor large temperature swings, which add cost and potential scrap.

7.4   Bolt-In

The term bolt-in mold for TSE molding generally refers to a two-plate compression mold that is secured to the platens, and as the press opens the operator reaches in, demolds, cleans, and loads material.

7.5   Shuttling

Cure times and change times are relatively lengthy in TSE molding. Couple this with having to load inserts and/or preps, and shuttle presses become an economical advantage over conventional molding. Shuttle presses incorporate either two entire molds, or one mold with two shuttle plates. The principle is that as a mold is curing in the press, the other mold, or mold plate, is being demolded, cleaned, and loaded to be ready to shuttle into the press. Figure 7.3 compares a typical book mold sequence to a shuttle sequence. Clearly, there is an overall cycle time advantage to shuttling in that many operations can be performed in tandem with the cure cycle.

7.5.1   Double Shuttle

Two complete molds are used in a double shuttle press. While one mold is in the press during the cure cycle, the other mold is shuttled to its respective staging station. Figure 7.4 shows a schematic of the potential stations in a double shuttle. Imagine a mold being shuttled to the right staging station. In tandem, a duplicate mold is shuttled into the press. The press closes and initiates the cure cycle. In the meantime, in the right staging station, the other mold is opened, and in this case, one of the plates is flipped to the right K.O. station where the molded articles are removed. The plates are cleaned and inserts are loaded (if required), the mold is closed and ready to be shuttled into the press. Once the duplicate mold completes the cure cycle, the press opens, the mold shuttles to the left staging station, and now the mold that just completed its mold-open cycle is shuttled into the press for its cure cycle.

Most shuttle presses as described move molds from station to station while bolted to their respective heated platens. This allows the mold to maintain a source of heat throughout the molding cycle. Unfortunately, this is usually only done with the bottom platen. The only time the top plate of the mold reaches its heat source is when it is clamped in the press. Otherwise, all heat is transferred from the bottom platen. Double shuttle systems can be incorporated into compression, injection, or transfer molding, or any derivative thereof.

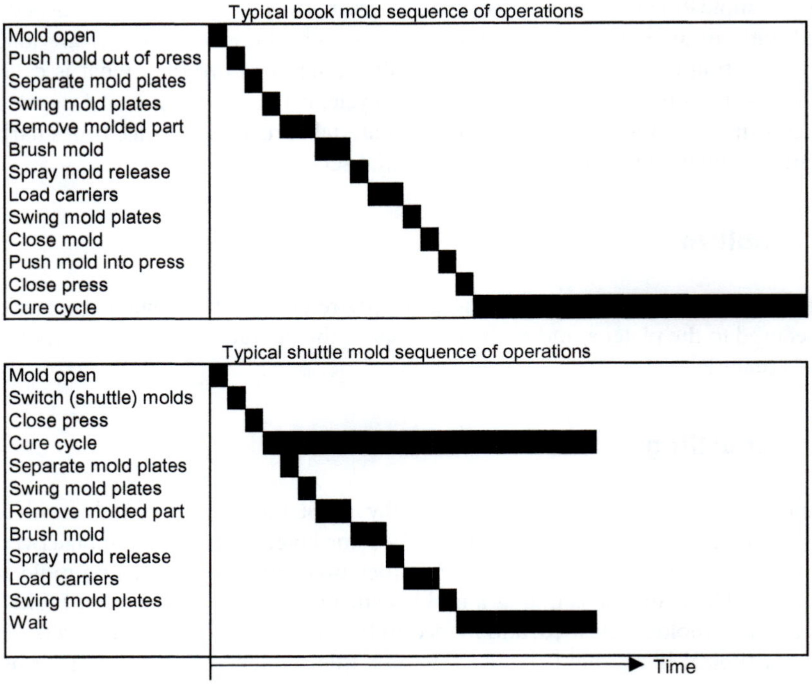

Figure 7.3 Shuttle mold sequence comparison

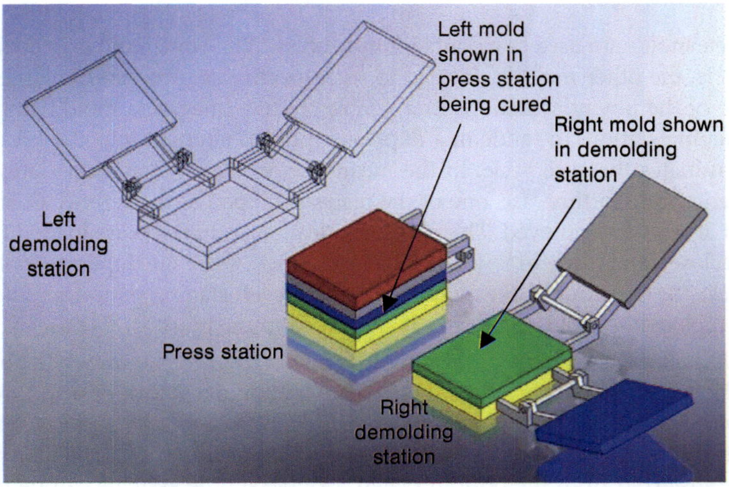

Figure 7.4 Double shuttle layout

7.5.2   Single Plate Shuttle

As mentioned earlier, the longer the mold-open time, the greater the heat loss. Single plate shuttle takes advantage of a very short mold-open time. The mold is open long enough to shuttle out a plate housing the molded part and exchange it with a sister plate that is cleaned, demolded and charged with metals (if required). An auxiliary heating platen can be incorporated at the K.O. station to assure the sister plate will not drop in temperature as it is staged, waiting to be shuttled (see Figs. 7.5 and 7.6). This heating of the sister plate is an important step not to be overlooked. In essence, the sister plate can be heated to the exact temperature required in the mold, resulting in virtually no heat loss during the mold-open cycle. In contrast, without the input of heat, shuttling or book-type molds lose heat drastically and need to reach the mold temperature during the cure cycle when the mold is closed. This extends the cure time. Therefore, a heated shuttle plate not only reduces mold-open time, but also the cure time, resulting in extremely short cycle times and avoiding scrap created by temperature fluctuations.

In single plate shuttle, top and bottom mold plates will be bolted directly to the heating platens. In constant contact with the platens, the mold will maintain an even temperature and will not be as heavily influenced by variation in mold-open time. Single plate shuttle

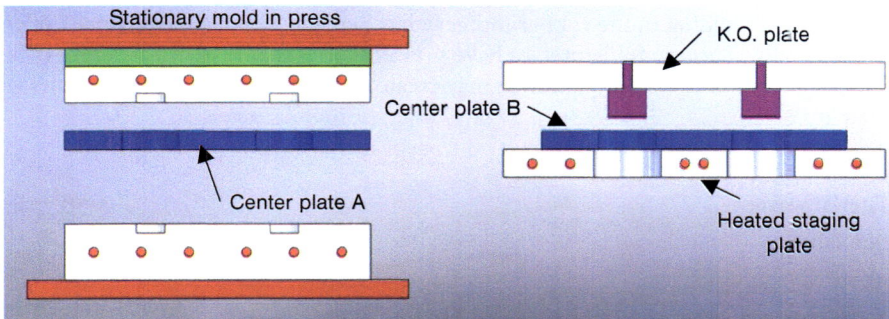

Figure 7.5 Single plate shuttle

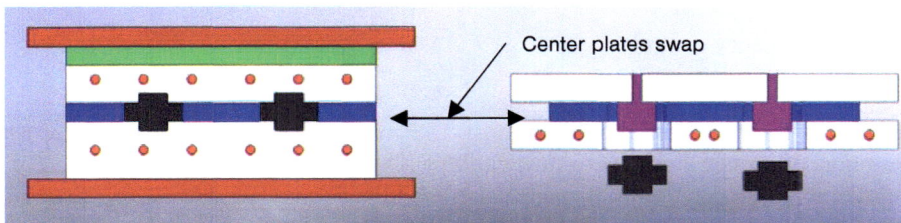

Figure 7.6 Single plate shuttle

systems will achieve greatest success with products having one or several of the following characteristics:

- High volume
- Flashless mold design
- Parts that get captured in a center plate during demolding
- Materials susceptible to mold temperature fluctuations
- Parts requiring a lengthy mold open time

Shuttle systems can reduce labor, scrap, and cycle time. Injection shuttle systems coupled with flashless molding techniques offer an excellent foundation for automation. Cycle time is reduced because demolding does not adversely affect the cycle time. To take complete advantage of single plate shuttle, all mold-open operations must be contained on the plates being shuttled. If brushing of all mold plates, or removal of a runner system is required, it diminishes the advantages of single plate shuttle. Therefore, if used with injection, the runner, sprues, cold drop remnant, etc., must all be captured in the shuttle plate.

7.6   Rotary Molding

High-cavitation molding in the rubber industry has become commonplace since cycle/cure times are lengthy. Larger molds/presses have prevailed in a cost system that argues "bigger is better". Unfortunately, larger platforms may create a host of process inconsistencies which translate to:

- High scrap
- Material waste
- High labor
- Expensive equipment
- Expensive tooling
- Lengthy mold cleaning time
- Large secondary equipment to handle larger platforms
- Temperature variation
- Difficult inventory control
- High mold-change time

Instead of a batch process typical of most TSE multi-cavity molding operations, rotary molding incorporates a one-piece flow concept. The idea of rotary molding is to concentrate the molding operation to a single cavity mold (or some lower cavitation than would be expected in a typical batch process). It is difficult to dispute that a single cavity mold is easier to process than a multi-cavity mold; or stated differently, more cavities in a mold offer greater variability in processing.

Most rotary molding machines have stations (or pods) that have their own clamp cylinder (see Fig. 7.7). Hydraulic pumps can be mounted directly to the rotary dial to avoid complicated and maintenance-intensive rotary unions.

The rotary machine depicted in Fig. 7.7 is an injection molding machine. However, rotary machines can also be used for compression or transfer molding. In the case of compression or transfer molding, a material delivery system needs to be in place such as a pick-in-place robot or an operator delivering material manually.

Additional rotary dials and/or feeder bowls adjoining the molding dial can be incorporated to perform bonded carrier loading, trimming, assembly, inspection, or other operations that can automate a rotary molding process.

Quick mold change capabilities should be designed into each station to allow for easy cleaning and maintenance. Spare molds allow for the process to continue running while maintenance or cleaning is being performed. In fact, some machines can be programmed to keep track of individual molds and sound an alarm when predetermined cycles are attained for maintenance or cleaning.

The 12-station rotary table shown in Fig. 7.8 shows five active stations: open, de-mold, load, close, and inject. In addition, there are seven curing stations. As the table rotates these stations transition to the next operation. Consider a 15-second cycle time at each station; this translates to one part (if a single cavity mold is used in each station) every 15 seconds. Total cure time would be $15 \times 7 = 105$ seconds.

Similar to the shuttle processes described earlier, the molds in a rotary press are bolted directly to their own heating platens. Therefore, heat loss is minimized and a reduction in cure time may be realized in rotary molding. A raw comparison between a rotary process versus a

Figure 7.7 Twelve-station rotary injection molding machine (Courtesy of DESMA [2])

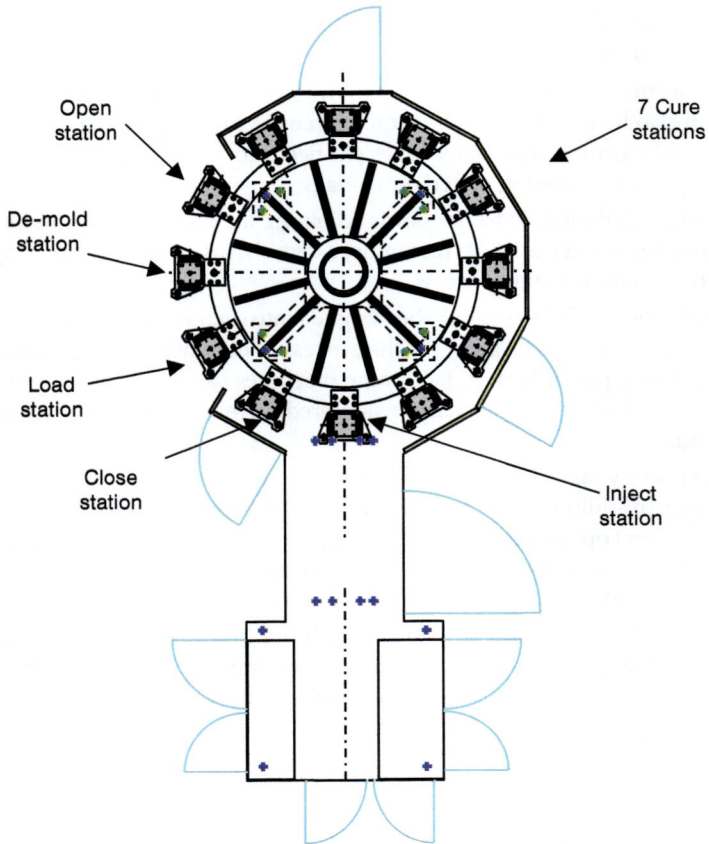

Figure 7.8 Twelve-station rotary injection molding machine layout (Courtesy of DESMA [2])

conventional batch process is given in the following. For ease of comparison, a 105 second cure time is assumed for both processes.

Conventional multi-cavity batch process	Rotary molding
6 s mold close	15 s each station
105 s cure	
6 s mold open	
45 s part removal	
30 s load	
192 s total cycle	

Using a 9-cavity mold for the multi-cavity process and a 12-station rotary machine, the following productivity rates are achieved:

Multi-cavity process $3600/192 \times 9 = 168$ pcs/hr

Rotary process $3600/15 = 240$ pcs/hr

Assuming a 16-cavity mold instead of 9 cavities for the multi-cavity process, the added cavities yield 300 pcs/hr. So, a rotary process does not always yield more parts per hour. To justify a rotary machine other factors need to be considered:

- Capital investment
- Tooling investment
- Scrap savings of rotary over conventional molding
- Potential cure time reduction since molds are bolted directly to a heat source
- Ease of automation (can direct labor be removed)
- Capacity
- Press efficiency (rotary machine can run continuously through mold cleaning)
- Machine rate comparison ($/hour)

Rotary molding can be a very efficient molding method. The advantages of one-piece flow are difficult to argue. Careful evaluation of costs, along with the actual anticipated advantages, need to be weighed before considering rotary molding.

7.7   Core Bar

Some parts such as convoluted bellows or spark plug boots, cannot be molded by conventional means. Normally, a round part would be molded with the mold parting line diametrically and parallel to the press parting line. Since a bellows boot mold could never be machined in this fashion, the part is laid on its side and the mold parting line runs perpendicular to its diameter. Spark plug boots are molded similarly. This molding method is sometimes referred to as a tree mold, because a long bar holds many cores, usually on both sides, and resembles a tree. Core bar molding lends itself well to the injection molding process, but it can be used with transfer or compression molding as well.

Demolding of bellows boots usually involves air which inflates the molded bellows off of the core and allows the part to slide over the convolutes. Often fixtures are used to capture the ends of bellows in such a way that the air tightens the molded bellows onto the fixture as it inflates over the core. Spark plug boots usually use a mechanical scraper to strip the boot off of the angled pin.

Rotating core bars, or double sets of core bars are common for this type of molding. As one set is curing, the other set of cores is being demolded.

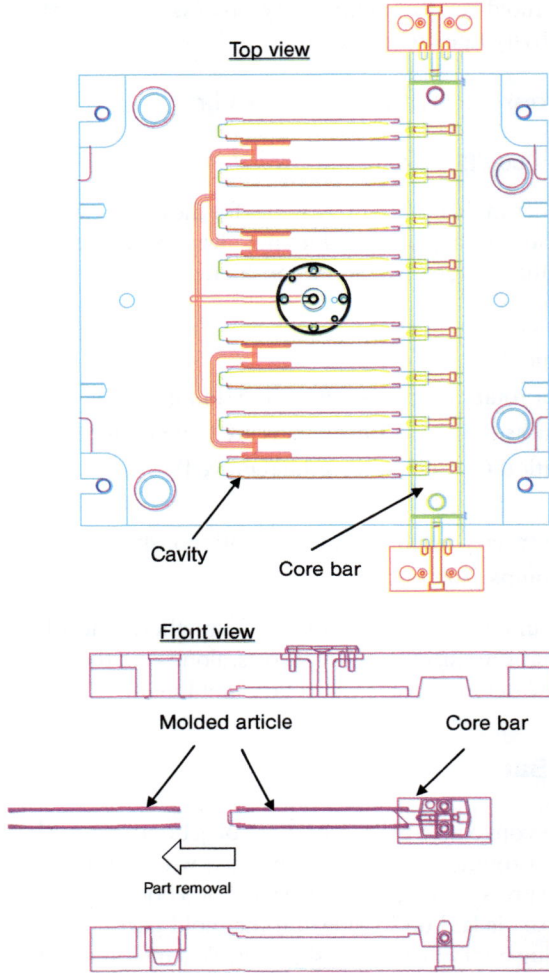

Figure 7.9 Core bar mold with 2 core bars (Courtesy Mid-States Tool and Machine, Inc. [3])

7.8   Mold Construction

Molds that have the cavity machined directly into the mold plates are referred to as cut-in-plate. A simple two-plate mold with a cavity machined into one or both plates can suffice in many applications. These molds can be constructed from 4140 grade steel and will hold up well for many years, provided that the final part does not have any critical features or tolerances.

Molds can also be built with inserted cavities. In this case, plates — usually 4140 steel — house a series of cavities that can be manufactured with hardened steel or stainless steel. There are several reasons for inserting a mold, the least of which is cost. Inserted cavities can be self-registered. Round cavities can easily be self-registered by telescoping close-fitting diameters into the opposite half of the mold. It is important to note that all, or all but one, insert must allow adequate clearance in the mold holding plates to float in an effort to align with the opposite half, or binding will occur.

Self-registration can be accomplished with pins, tongue and groove, or other means. In any case, self-registered molds offer consistent registration between cavity halves, which is useful when tight tolerances are required for the final product.

Inserted cavity molds offer the advantage of easier mold repair. If spare cavities are on-hand, replacing a damaged cavity with a spare can be accomplished in minutes, whereas cut-in-plate can take weeks to repair. Inserts should be designed to allow for cavity exchange without removing the entire mold from the press. This requires that snap rings or other retainers have access from the mold parting line.

7.8.1   Cavitation

Cavity spacing can be a difficult topic to discuss in standard terms because of the multitude of mold constructions and part geometries that exist. Self-registered inserted molds will require larger cavity spacing than cut-in-plate configurations. Attention should be given to the webbing thickness of holding plates. Particular consideration should be given to this thickness in molds that require high knockout, or stripping forces, and/or when holding plates are substantially thin.

Cavity patterns should be honeycombed to allow as many cavities as possible in a given area. Cavities should be kept within the diameter of the press ram. Therefore, round patterns are optimal. Square patterns with the corner cavities outside the ram diameter are susceptible to thickness variation and molding defects. Note the two cavity layouts in Fig. 7.10; if all cavities

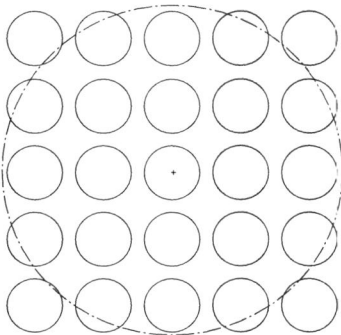

Figure 7.10a Square cavity pattern

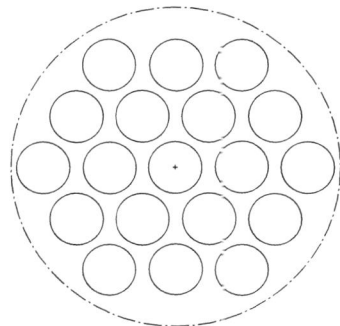

Figure 7.10b Round cavity pattern

were required to fit within the outer ring (press ram, or pot diameter), the square pattern would only yield 13 cavities, while the honeycomb patterns yields 19.

7.9   Article Removal/Ejection

Ejector pins as used in the plastic injection molding industry are useless in TSE molding. First, hot TSE parts are elastic and ejector pins can nudge a molded part away from a moving ejector pin, but may not completely dislodge it from the cavity. Second, and more important, ejector pins cannot be machined with a tight enough clearance to avoid TSE leakage down the pin, and still be considered a working fit. Many TSEs flash at 0.0002″ gap. Therefore, conventional straight K. O. pins will not work.

Valve angle fit K. O. pins can be used effectively in many cases. Cavity pressure forces the valve angle shut and prevents TSE from leaking under the pin. An air blast under the pin can assist in removing the part and to keep the valve angle clean of debris and flash.

Stripper plates are an effective method of removing parts from a core. Round parts work well with stripper plates. A good valve angle fit is essential to keep TSE from bleeding. Stripper plates in multi-cavity molds should house separate inserts (stripper inserts) that are allowed to float in the plate and to be able to register independently onto their respective cavity inserts. Stripper insert parting lines need to be positioned in a robust portion of the part to withstand ejection without tearing the molded article. If bonded inserts are used, locating the parting line under the bonded insert transfers the force to the bonded insert instead of the more delicate TSE.

Compressed air can be the simplest method for removing parts from a mold. An air hose can be directed at the molded article to blow it off of the mold. This method can lead to lost parts and contamination, if parts are not appropriately captured during the air blow-off. Vacuum is an efficient way of removing parts and flash from a mold. An industrial vacuum with a large diameter non-corrugated hose can remove articles from the mold and capture them in a filtered basket. This method makes for a clean demolding operation.

7.10   Mold Cavity Finish

The finish on a mold cavity can play an important role in how the TSE processes during the molding operation and how the molded article works in the final application. In most cases, a matte finish is desired on the mold cavity surface. A diamond polished, mirror finish on a mold may look impressive, but most TSEs tend to stick to very smooth surfaces — particularly softer durometers. A vacuum is created when a very smooth TSE surfaces mates with a mirror finish in a mold cavity, resulting in high part removal forces that may cause a tear.

A fairly rough mold cavity surface actually assists the flow of many TSEs. As the TSE travels across the cavity surface, a rough topography will cause a tumbling effect at the flow front which prevents skinning.

In most TSE applications, it is recommended to highly polish mold cavity areas to remove any tooling marks or grooves. Once all of the tool marks are removed, the cavity should be

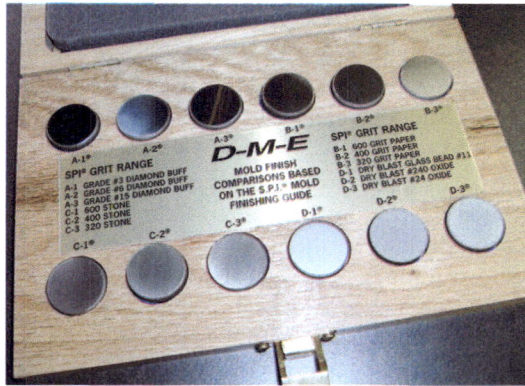

Figure 7.11 SPI mold surface finish guide

blasted in a sand blaster with aluminum oxide media. Do not assume that the aluminum oxide blasting will remove tool marks. These tool marks may not be clearly visible after blasting, but under magnification may reveal grooves that can cause parts to stick or tear during removal from the mold.

Quantifying a cavity surface finish leaves much to interpretation. However, experienced mold builders can replicate finishes by knowing what media creates certain finishes. Some standards do exist that designate alphanumeric values to degrees of coarseness. The Society of the Plastics Industry, Inc. (SPI) has created such standards. Figure 7.11 shows sample inserts with various finishes. These samples should be used to visually compare if the mold cavity meets the designated criteria. Considering the SPI standards, typical TSE molds should have a D-2 (dry blast #240 aluminum oxide) or D-3 (dry blast #24 aluminum oxide) finish.

For many TSE seal applications a matte finish is desirable because fluids tend to penetrate the tiny pockets created by the rougher surfaces and assist in lubricating the seal. Also, matte finishes on the molded article cover up flow lines, discolorations, fingerprints indentations due to parts touching each other, and surface anomalies that do not affect fit, form or function. Highly polished mold surfaces may produce more rejects simply because non-critical visual characteristics become more obvious.

It should be noted that molded articles do not necessarily conform exactly to the mold cavity surface finish. Polymer types, color, molecular weight, filler types, molding method and even cavity pressures may prevent the TSE from completely penetrating the pores of the mold cavity surface and replicating the desired surface finish. Trials should be performed during prototyping to establish the desired finish on the molded article.

7.11   Heaters

All TSE molding methods require heat to cross-link the end product. Heat is generally supplied by the use of electric heating rods (or cartridge heaters), although steam and hot oil are still used by some molders. Autoclaves and microwaves can also be used to crosslink TSEs, but will not be covered in this text. Discussions will concentrate on cartridge heaters.

What is important when heating molds is that the mold needs to maintain a constant temperature throughout the entire molding area. This logical statement is much easier said than done. All too often having an adequate supply of heat is not given the forethought necessary.

Cartridge heaters should have sufficient wattage to quickly recover heat loss due to mold-open operations that allow heat to escape from the mold surfaces. This can be caused by air blow-off, mold lubes, fans or open windows in the molding area and complications in demolding resulting in longer mold-open times, or others. Steps to avoid heat loss should be instituted where possible. The use of compressed air should be minimized. Insulator boards should be mounted to the outside edges of molds to capture heat in the steel.

TSE molding typically incorporates cartridge heaters into the molding platens as opposed to the mold itself. This is due to the fact that in TSE molding, book-type molds are still a preferred method. Book-type molds are removed from the press and plates are swung in various directions to demold and clean the parting lines. In many cases, plates are very thin, and to adequately vent molds, numerous parting lines are incorporated making it difficult to provide heaters to the mold plates. The wire-to-heater interface in a typical cartridge heater is fragile and can easily break the heater insulation if repeatedly fatigued, resulting in a failed heater. Therefore, heaters are best used in platens or in molds that are not removed from the press for demolding. In addition, mold cleaning of TSEs typically involves submerging a mold in a cleaning solution, or harsh grit blasting, which would severely limit the life of heating rods.

Where space permits, large diameter cartridge heating rods should be used. Heaters should be equally spaced. Since heat dissipates more at edges and corners of molds or platens, it is a good idea to use heaters that are wound to generate more heat on the ends than in the middle. In addition, zoned heating should be incorporated to apply more heat to areas that have a tendency to drop in temperature. Thermocouples need to be placed in a location that can monitor or simulate mold surface temperature for a specific zone and not read and affect an adjacent zone.

7.11.1   Heater Calculation [4]

A simple rule of thumb for calculating wattage for a steel mold is to use 6 watts/in^3 of steel. However, to more accurately calculate actual wattage requirements, the following formulas can be used which take into account start-up and running requirements.

Start-Up

Start-up provides energy to heat a mold in a targeted amount of time plus the energy losses though convection and radiation plus a 20 % safety factor.

Parameters

Specific heat of steel (c_p) (0.11 BTU/lb · °F)   116.05 J/lb · °F
Density of steel   0.28 lb/in^3
Heat loss via radiation and convection of oxidized steel   2 W/in^2 at 400 °F
                                                            1.2 W/in^2 at 300 °F

Heat loss via conduction (k)
        Typical insul. board 1.90 BTU/(ft^2 · in · hr · °F)   0.56 (Watt · in)/(ft^2 · °F)

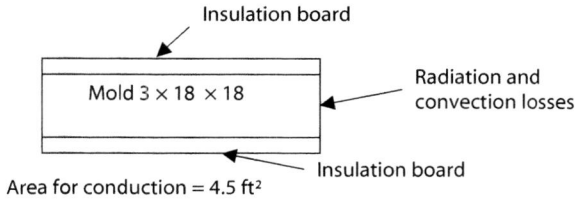

Insulation board

Mold 3 × 18 × 18

Radiation and convection losses

Insulation board

Area for conduction = 4.5 ft²

Example

18″ × 18″ × 3″ steel mold
60 minute target time for start-up
Final temperature 400 °F
Weight of mold is 272 lbs

Calculation

Energy to heat up mold c_p × wt. × ΔT/3600 s
                          116.05 × 272 × 330/3600 = **2894 Watt**

Losses
    Radiation and convection   2 W/in^2 × area radiating
                                2 × 18 × 3 × 4 = **432 Watt**

    Conduction   area (ft^2) × k × ΔT/thickness of insulation
                                4.5 × 0.56 × 330/(1/2) = **1663 Watt**

2/3 factor   Watt loss × 2/3
                                (432 + 1663) × 2/3 = **1395 Watt**

Total   1395 + 2894 = **4289 Watt**
Plus 20 % safety factor   4289 + 20 % = **5147 Watt**

Using the 6 watts/in^3 18 × 18 × 3 × 6 = **5832 Watt**

Note: The conduction, and radiation and convection losses are calculated as though the temperature differences are always 330 °F. In actuality they are not — they are gradually building up. The ⅔ factor accounts for this difference.

7.11.2   Running

When a mold is running, the heaters only need to supply heat to offset losses through conduction, and radiation and convection heat to process and cure the material. In this case, the ⅔ factor does not apply to conduction, and radiation and convection losses, since the mold is effectively at temperature all the time. Also, the radiation and convection losses apply to a larger area for part of the cycle, when the mold is open.

Assuming the same mold as described earlier with a mold-open time of ⅓ the total cycle, the following formulas apply:

Radiation and convection losses	2 W/in² × area radiating × (% open)
	432 × 1.33 = **575 Watt**
Conduction	area (ft²) × k × ΔT/thickness of insulation
	4.5 × 0.56 × 330/(1/2) = **1663 Watt**
Total	575 + 1663 = **2238 Watt**
	Plus heat to process and cure material

7.11.3   Conclusion

Unless the amount of energy needed to maintain a mold temperature is necessary for power consumption information, the running calculation may not be needed. The start-up energy calculation clearly shows a higher wattage requirement, and should be used to determine the correct heaters in the construction of a TSE mold. The 6 watts/in³ closely resembles the long version calculation with an added safety margin. Unless complicated thermal dynamics are involved, the 6 watts/in³ rule would suffice in most applications.

7.12   Heat Transfer

Materials with high heat conductivity can be used to draw heat to dead zones or tall profiles in molds to minimize mold heat variation and reduce molding cure time. Unfortunately, most materials with a high coefficient of heat transfer are relatively soft. However, pockets can be machined under cavity surfaces; hollowing out sections where these softer materials, such as copper or bronze alloys, can be inserted to transfer heat and have no adverse effect on the mold integrity. Some copper alloys, such as Ampcoloy 944, are hard enough to withstand TSE molding and cleaning, and along with hard chrome protection can be used directly in the mold cavity provided that they are not prone to damage from insert molding or unusual

Table 7.2 Thermal conductivity of mold cavity metals (Courtesy of Ampco Metals S. A.) [5]

Mold Material	Thermal Conductivity W/m·°K	Tensile Strength MPa	Hardness Rc
Ampcoloy® 944*	156	938	31
Beryllium copper (2 % Be)	105–130	1100–1300	N/A
Stainless steel (420)	36	862–1724	58
Tool steel (P20)	38	900–1000	30
Tool steel (H13)	24	1420	48

*Ampcoloy® is a registered trade mark of Ampco Metals S. A.

stresses. Table 7.2 shows hardness, tensile strength, and thermal conductivity for various mold related steels, beryllium copper and Ampcoloy® 944.

Some copper alloys too soft to be used for mold cavity surfaces have thermal conductivity properties exceeding 350 W/m·°K which make great heat conductors in close proximity to mold cavity surfaces.

Other effective means of transferring heat are heat pipes.

"Isobar® Heat Pipes are super-thermal conductors that have the capacity to transfer large amounts of heat at high speeds in both heating and cooling applications. In fact, Isobar® Heat Pipes have, in some applications and orientations, a thermal conductivity in excess of 20 000 times the rate of a solid copper bar of the same geometry. Isobar® Heat Pipes are also high-speed superthermal distributors. The working fluid constantly changes phase due to the low vapor pressure inside the unit — the device is innately isothermal. Temperature uniformity along the entire length from the evaporator to the condenser is typically in the range of ±1 °C while transferring large amounts of energy". [6]

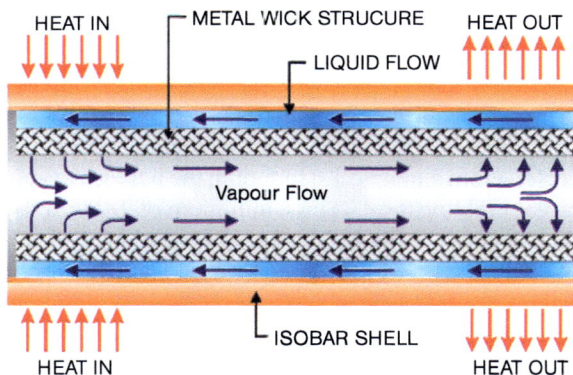

Figure 7.12 Isobar® heat pipe explained (Courtesy of Acrolabs Ltd.) [6]

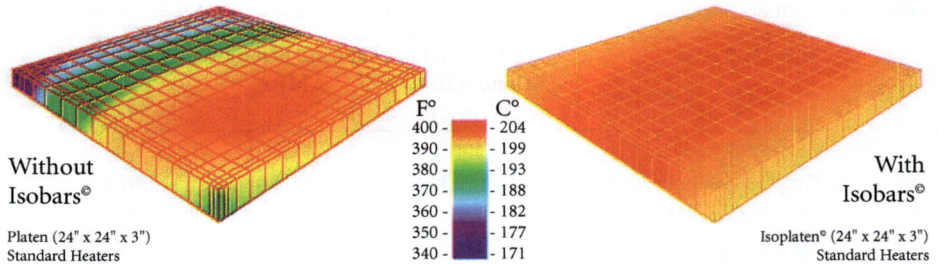

F°	C°
400 -	- 204
390 -	- 199
380 -	- 193
370 -	- 188
360 -	- 182
350 -	- 177
340 -	- 171

Without
Isobars©

Platen (24" x 24" x 3")
Standard Heaters

With
Isobars©

Isoplaten© (24" x 24" x 3")
Standard Heaters

Figure 7.13 Isobar® heat pipes installed in a platen (Courtesy of Acrolabs Ltd.) [6]

Figure 7.12 shows a cross-section view of an Isobar® heat pipe and how it transfers heat. Heat pipes can also be used in platens to equalize temperatures. One method of utilizing heat pipes in a platen is to install heater rods in one direction and install heat pipes perpendicularly at a level closer to the mold (shown in Fig. 7.13).

7.13   Insulation

Insulation boards are used between the heating platens and the presses bolster plate so that the press does not become a heat soak. The same boards can be used to insulate portions of the mold to maintain temperatures safely below the TSEs crosslink point such as cold runner systems or wasteless transfer molding. These boards should be of good quality with high compressive strength. Insulation boards have limited service life and should be replaced on a routine preventative maintenance program. If the insulation boards become soaked with liquids, the insulation properties as well as the compression strength drops dramatically. In the case of a liquid spill, the boards should be replaced.

7.14   Vacuum

Out-gassing is a major concern in the TSE molding process. Drawing a vacuum in the mold to expel gases has proven to reduce molding defects such as air-traps and sponge. A perfect vacuum is considered impossible to attain. Perfect vacuum would suggest an atmosphere containing no molecules. From a practical standpoint, the best achievable vacuum is 29.5 Hg (inches of mercury). For molding most TSEs, a minimum of 27 Hg is required to adequately evacuate the air in a mold cavity to minimize air traps and sponge. Therefore, a vacuum gage should be located in an area visible to the operator (see Fig. 7.14c). In fact, vacuum should be electronically monitored and alarms should be instituted when the vacuum falls below 27 inches of mercury.

Parting line vacuum is a common method for evacuating entrapped air; however, it has some drawbacks. Parting line vacuum attempts to draw entrapped air once the mold is closed. A hole is drilled near the cavity inboard of an O-ring, which is placed along the parting line

Figure 7.14a Vacuum box front view

Figure 7.14b Vacuum box view underneath [7]

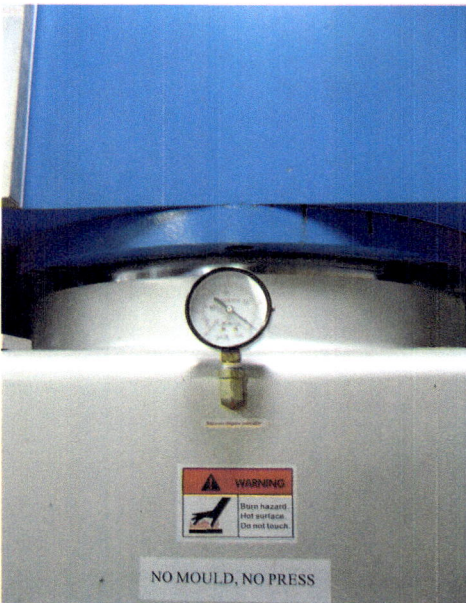

Figure 7.14c Vacuum box with vacuum gauge (Courtesy of Fullwell Holdings) [7]

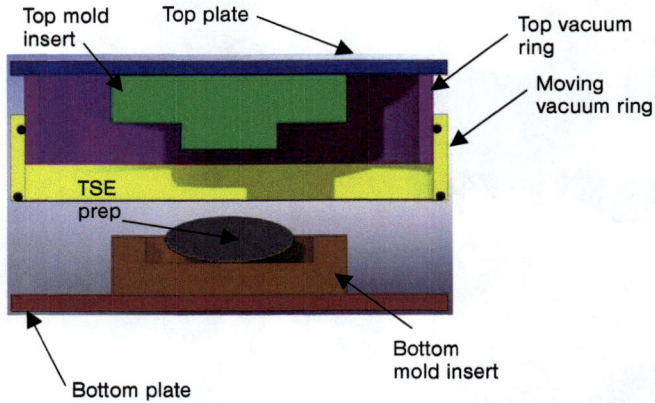

Figure 7.15a Vacuum box open

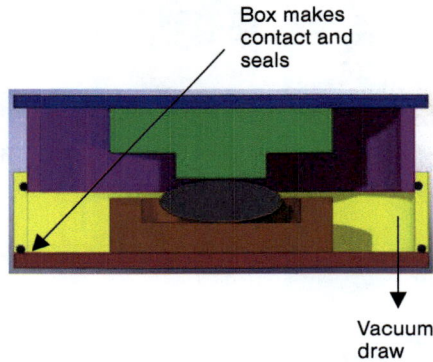

Figure 7.15b Vacuum box contact

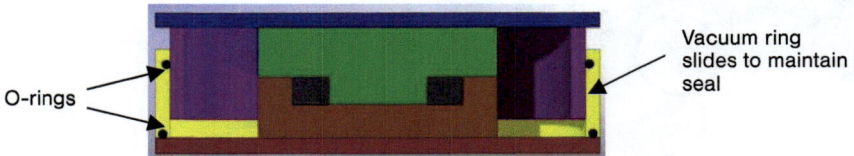

Figure 7.15c Vacuum box closed

of the mold around the periphery of the cavity area. The O-ring attempts to seal the outside air from being drawn into the cavity. There are several disadvantages to parting line vacuum:

- Inadequate volume of air draw — cavity spacing, size, and number generally optimize the available mold size. Therefore, limited space is available to include vacuum holes. Drawing the volume of air required to evacuate the cavity takes time. Having small vacuum lines only increases the amount of time required to achieve full vacuum.

- Air trapped in the cavity once the mold is closed — even if large holes are used to draw vacuum from the cavity, the mold needs to be closed to effectively seal minimizing the effect of air escapement. Since ideally the mold cavity should have a tight shut-off with the other half of the mold, air is not allowed to be drawn out. Typically, vents are cut into the mold at strategic locations (usually the last place for the cavity to fill) at the periphery of the cavity that feed to larger passages and ultimately to the vacuum holes.
- Vacuum lines plugged with TSE — overfilling the cavity can transport TSE into the vacuum holes blocking the source of vacuum to the cavity. Also, flash removed from the mold during the mold-open cycle can inadvertently find its way to the vacuum hole and plug the hole.

Vacuum boxes are by far the preferred method of evacuating air from the cavity. Many iterations of vacuum boxes exist, but they share the same premise. As the press closes, a movable portion of the box (generally an outer ring) is allowed to make contact with the opposite half of the mold or platen and create an air-tight seal via O-rings. This seal is accomplished before the mold contained inside the box makes contact with the other half. A valve opens and draws vacuum through a very large opening that is plumbed to large accumulators. The idea is that the large plumbing and accumulators can remove the air from the inside of the box very quickly. Once the air is removed, the press continues to close. The outer ring is allowed to slide along an inner ring to maintain a vacuum seal as the mold closes (see Fig. 7.15a–c). Vacuum boxes are used successfully in all molding types, although they become difficult to incorporate with automation and plate handling methods because the box tends to get in the way of press and plate movements.

As mentioned above, vacuum boxes are very effective, but cumbersome. An effective hybrid of parting line vacuum is a delayed-close parting line vacuum. Here, a taller profile seal is used along the parting line. This seal makes contact with the opposite half of the mold before complete mold closure occurs (perhaps 0.030″ mold-open). The closure is delayed and vacuum is activated. Once full vacuum occurs, the mold is allowed to complete its closure cycle (see Fig. 7.16). This method is considered a "poor man's vacuum box". It draws vacuum in an open mold similar to a vacuum box, but does not have the complications of hardware hanging from the top half of the press. However, keep in mind that the seal used in this technique is susceptible to damage, and therefore requires strict attention. Vacuum boxes are

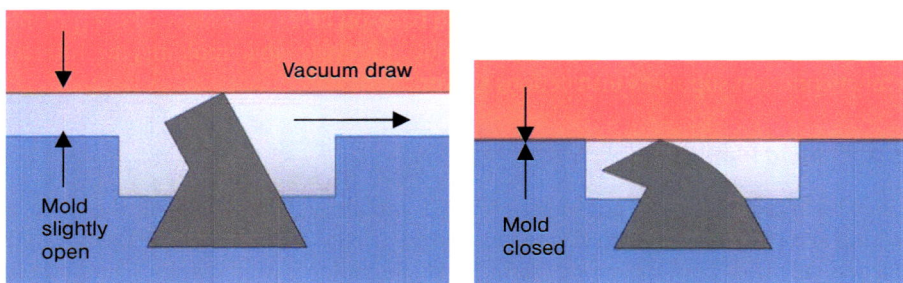

Figure 7.16 Parting line vacuum with open mold draw

difficult to incorporate into injection molding process. Vacuum boxes are best when used in a vertical press platform since the housing is very heavy and the weight can influence binding and premature wear. Therefore, parting line profile seals are quite effective in injection molding.

7.15   Release Aids

7.15.1   Mold Lubes

A variety of spray-on mold releases are available. These mold lubes are designed to offer lubrication to the mold cavity surface. Not only do these lubes aid in molded article removal, but they also may assist in filling the cavity. As much as it may be tempting to use lubes, they should be used only after conventional processing and mold plating options are exhausted. Using mold lube can create greater problems than simply helping with article removal. For example, spraying is not exact. Every operator has a different technique and uses different quantities of lube. As explained repeatedly, variation in TSE molding causes scrap. Mold lubes alter the flow of the TSE. In addition, mold lube is technically a contaminant. As flow fronts come together, they carry anything that is loose in the flow path including fluids. The very nature of mold lubes is to keep things from sticking together. Flow fronts are exactly what should not be kept from knitting together. Mold lubes create knit lines that are typically the weakest portion of the molded article. Also, mold lubes can hinder in-mold bonding as flow fronts meet the substrate to be bonded and a barrier is formed by the mold lube between the TSE and the adhesive on the carrier to be bonded. Lubes diluted with water tend to be overused and can cool the mold surface substantially creating even greater molding variation.

Careful attention should be given to supplier instructions. Pay particular attention to ingredients and MSDS sheets. Some molded products stipulate a list of chemicals that must not come in contact with the end product. Mold releases are usually designed to work in tandem with specific TSE base polymers. They are available in aerosol spray cans, or they need to be diluted with water and sprayed with a pneumatically charged spray gun. The instructions for applying mold releases vary from spraying the mold cavity surfaces between molding heats, to something as elaborate as spraying the mold cavity when it is cold, allowing it to bake while the mold heats up, spraying a different lube once the mold reaches operating temperature and allowing that to bake on for a few minutes, and then using another lube periodically between heats as needed. Although bake-on lubes do work well and potentially have less contamination than lubes that are applied regularly while molding, they do require careful attention to supplier's instructions to work properly.

7.15.2   Mold Plating

Platings for molds offer two advantages for TSE molding. Platings protect the mold from corrosion and wear, but they can also reduce surface friction and assist in part removal. Traditional and still most popular among platings for TSE molds is hard chrome. Hard chrome is an excellent mold protection, has normal conformability to textures, and offers some release aid. It should be noted that in many critical medical molding applications, chrome is not allowed. In the event that chrome may chip or flake from the mold and be carried by the molded article and potentially be released into the blood stream, it is best to avoid the use of chrome altogether.

A word about platers: finding a reliable and consistent plater is difficult. Many platers plate a variety of materials and products, of which some do not require the integrity needed in TSE molds. Choosing a plater should come from experience or field recommendations. It is wise to audit the plater's facilities and pay particular attention to cleanliness and controls that monitor their process. Of even greater concern is how mold components are handled. Particular care should be taken, if the plater routinely plates bulk packaged parts. Most complaints involving platers is their disregard for the delicate nature of mold components. Handling mold inserts and making sure they are segregated — not allowed to contact each other — is often the exception at platers. For this reason some mold builders have incorporated their own plating line to assure a high quality application.

A growing body of evidence suggests that specialty platings/coatings can increase productivity through reduced cycle times and less scrap. These platings can be extremely useful in specialty applications where sticky, corrosive, or abrasive compounds may be encountered. The best option is to try these coatings during prototyping or in a lab to determine compatibility with compounds, adhesives, surface finishes, or mold constructions. Some coatings get attacked by certain TSE curing agents and need to be ruled out. Others will require mild

Table 7.3 Various mold platings/coatings (Courtesy of Bales Mold Service, Inc.) [8]

Coating	Hardness Rc	Coeff. of Friction against Steel	Applied Temp. °F	Benefits
Hard Chrome	72	0.20 or less	130	Good abrasion resistance
Electroless nickel	50	0.45 or less	185	Moderate abrasion and excellent corrosion resistance
Nickel-cobalt	62	2.24 or less	185	Good abrasion and corrosion resistance
Diamond-chrome	85+	0.15 or less	130	Excellent abrasion resistance
Nickel-PTFE	45	0.10 or less	185	Excellent corrosion resistance, high lubricity
Nickel-boron nitride	54–67	0.05	185	Excellent lubricity, high wear and corrosion resistance, higher heat resistance than PTFE based coatings, uniform deposit, easily strippable, no effect on thermal conductivity
Titanium nitride (TiN)	90	0.40	900–950	Excellent abrasion and corrosion resistance, lubricity

mold cleaning methods where brushing and blasting may not be appropriate. Table 7.3 shows characteristics of common platings used for TSE molds.

A few cautionary comments regarding coatings/platings:

- Hard chrome is expensive. It may also require an extra conforming anode in complex geometry which may add cost. Chrome is also a carcinogen [9].
- Nickel-PTFE will require special maintenance precautions. Blasting molds (even with plastic media) will destroy this finish. Mold cleaning should be done with solutions known not to affect the coating. If in-mold bonding is done, metal inserts may remove this coating.
- Nickel-boron nitride has been known to be affected by certain sulfur bearing cure agents in TSEs. Trials should be done to determine material compatibilities.
- Electroless nickel is a low cost alternative to hard chrome and is also more uniform and can be used as a buildup of thickness under chrome [9]. However, it is softer and caution should be taken in harsh conditions. Electroless nickel may be a good cost alternative for holding plates that require corrosion protection.

7.16   Mold Cleaning

Molding TSEs creates residue on the mold surface. Regardless of the polymer being molded or the process chosen, crosslinking of TSEs leaves a residue. Within a few heating cycles, discoloration can be seen. Over time, the residue prevents venting, inhibits TSE flow, increases the tendency for trapped air, and if left on the parting lines, can alter part dimensions. Another form of mold buildup can be from adhesive wash, or flaking.

Mold releases can be applied to the mold surface to minimize mold fouling, but can also compound the fouling condition. Cleaning the mold with brushes between heating cycles can reduce buildup. Soft bristles such as nylon or tampico can be used frequently without compromising the mold surface. Wire brushes or abrasive fabrics should be avoided.

7.16.1   Plastic Media Blast

An aggressive approach to cleaning mold residue is to remove the mold from the press and place it into an enclosed media blasting machine. This machine uses similar principles as a sand-blasting machine, except small plastic pellets are used instead of sand or aluminum oxide. The plastic pellets are soft enough to avoid damage to the mold surface, but aggressive enough to physically remove stubborn residue, and strong enough to withstand repeated cycles of blasting. A variety of plastics are available in many different pellet sizes to suit particular applications. A drawback to plastic media blasting is that the mold needs to be removed from the press. A greater drawback is that, if inserted, the mold should be disassembled to remove all plastic media. This may require solvent cleaning inserts by hand. If not removed, plastic media left in the mold can damage parting lines and alter shut-off heights. Under no circumstances should aluminum oxide or sand be used to clean molds.

7.16.2   Ultrasonic Cleaners

Ultrasonic cleaning requires the mold to be removed from the press. The mold plates are separated and suspended in a bath of cleaning solution (usually an alkali solution) that is heated to approx. 190 °F. Ultrasonic waves, in conjunction with the supplier recommended cleaning solution, jar the residue from the mold surfaces. The mold is then placed in a second bath of clean rinse water. This process can be time consuming. It may take up to an hour to clean a mold. Ultrasonic cleaning is selective in its success. Some baked-on adhesives and crosslink byproducts are unaffected by ultrasonics and need a more aggressive method of removal. Ultrasonic suppliers can suggest settings and solutions that may suit a particular need. This method can be very useful in many applications and should be considered before more aggressive methods are used, which can shorten mold life.

7.16.3   Ice Blast

Dry ice blasting is an excellent way to clean molds without removing the mold from the press. The concept of this process is similar to sand blasting, except the media used is small particles of dry ice (frozen carbon dioxide). A block of dry ice is placed into a canister, and the apparatus shaves a layer of ice which is carried into the path of pressurized air and through a nozzle. The nozzle is pointed at a mold surface and the dry ice particles are cast with high pressure air against the mold surface to aggressively remove stubborn deposits. The ice particles quickly dissolve into the atmosphere with no residual harm to the environment and no clean up. Dry ice sublimates — transitions directly from a solid to a gas — at –109 °F and at a rate of five to ten pounds every 24 hours [10]. The fact that this cleaning method can be used without removing the mold from the press drastically reduces down time.

Drawbacks to this method include the fact that the unit is extremely loud and requires ear protection. It is cumbersome to use because the hoses are large, making the nozzle difficult to articulate, and can be ineffective in deep undercuts and blind pockets. Care must be taken along with adequate training to avoid injury with this apparatus. Accidentally contacting eyes, face, or skin can cause severe burns or major damage. Carbon dioxide is present in the atmosphere but can be harmful if inhaled in large percentages. Dry ice blasters should not be used in confined spaces or where there is poor air ventilation.

7.17   Conclusion

TSE molding has a multitude of options to consider for process optimization. Unlike plastic injection molding, where more standards exist in terms of presses and molds TSE custom molders must chart their own course through intimate knowledge of their own equipment and techniques.

References

1.   Fort Wayne Mold and Engineering, Fort Wayne, Indiana.

2.   DESMA, USA, Hebron, KY.

3.   Mid-States Tool and Machine, Inc., Decatur, IN.

4.   Hall, T. Phd., Personal interview 3/6/09.

5.   Ampco Metal S.A., Switzerland.

6.   Acrolabs Ltd, Windsor, Ontario.

7.   Fullwell International Holdings, LTD. Shatin, N.T., Hong Kong.

8.   Bales Mold Service, Inc., **www.balesmold.com**.

9.   Bales, S.J., Know your mold coatings, *Plastic Technology*, 12/04.

10. USGS, retrieved on 2/22/09 from **http://ga.water.usgs.gov**.

8 Compression Molding

Compression molding of TSEs can encompass a wide variety of techniques including the manufacture of tires. This chapter will focus on medium-to small-sized articles that typify the custom molded category.

Compression molding carries the stigma of an inferior or antiquated molding process. In some cases this distinction is deserved. However, as will be described, compression molds can incorporate complex internal movements that can produce flash-free, cost competitive articles. In addition, simple compression molds can produce low volume production quantities where tooling investment is a deterrent to launch.

In compression molding, typically a pre-weighed, cut-to-suit uncured prep is placed into a heated mold cavity. The press squeezes the uncured material to conform to the mold cavity. The mold remains under pressure and heat until the TSE is cured. The mold opens, and the cured article is removed. The TSE prep contains more volume than the finished product. As a result, excess material is allowed to flow into overflow areas, or tear beads, and is removed subsequent to molding.

8.1 Compression Presses

Compression molding was the first technique used to mold TSEs. Conceptually, it is the simplest molding method. Almost without exception, compression molding is performed with a vertical platform press. The press has an upper and a lower bolster, which usually houses the platens. The bolsters are held together with four posts (post press) or two plates

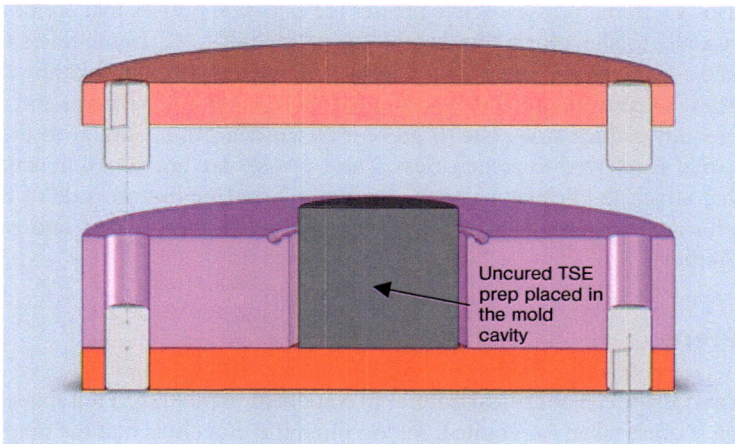

Uncured TSE prep placed in the mold cavity

Figure 8.1 Compression mold

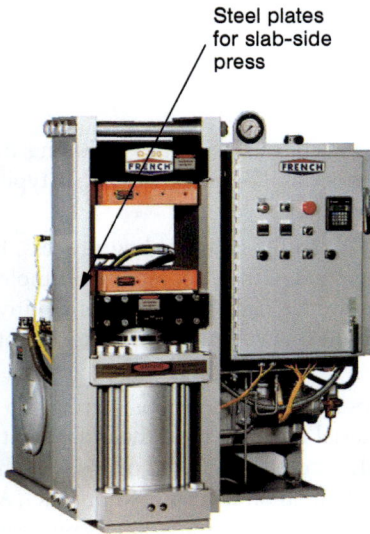

Figure 8.2a Slab-side press
(Courtesy of The French Oil Mill Machinery Co.) [1]

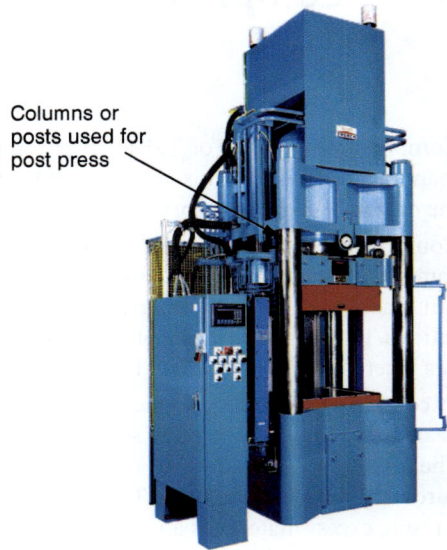

Figure 8.2b Post press

(slab-side press). A hydraulic cylinder (clamp cylinder) applies force to one of the bolsters and causes it to move in the direction of the opposite bolster in a closing fashion. In most cases, the clamp cylinder is under the bottom bolster. However, where the bottom platen height may be critical to other operations, down-acting presses (clamp cylinder is above the top bolster and forces the top bolster downward) are available.

There are also "C" frame presses. These presses are slab-side presses that are notched in the side plates, usually to allow for a shuttle table, or rotary table. "C" frame refers to the shape of the slab-side — the notch forms an opening to make the slab look like the shape of the letter "C". These presses offer plenty of clearance for secondary movements and automation, but usually require a thick side plate to prevent platen deflection. Similar to the "C" frame variation is what is referred to as tiebarless. These presses are horizontal injection molding machines that attach the bolster halves on only one side (the bottom side of the platens). This allows for ample room to include automation, part/runner pickers, and ease of mold installation/removal.

8.2   Preps

Prep weight and shape can be critical features in compression molding. Prep shape in compression molding is often taken for granted. All too often the thought is that the prep will simply conform during the process. Prep shape needs to be determined on a trial basis. The prep shape needs to be able to adequately fill the cavity and allow for out-gassing. Ironically, round

Figure 8.3 Cloverleaf shaped prep

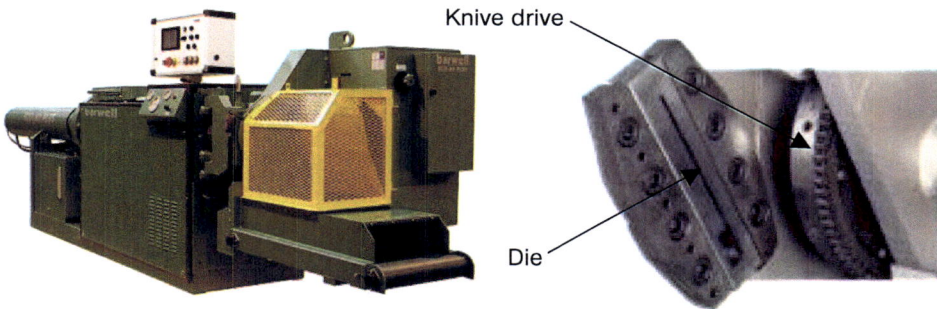

Figure 8.4a Prep cutter machine   **Figure 8.4b** Round cavity pattern

(Courtesy of Barwell) [2]

preps are not always best for round parts. Cloverleaf (Fig. 8.3) or pacman shaped preps are often better regarding non-fills or trapped air. Preps can be cut with shears or die-cut out of a slab of uncured TSE, or for higher volume applications an extruder with an integral knife blade can be used (see Fig. 8.4). Uncured TSE swells when extruded through a die, so die design needs to take this into consideration. Extruders with knife cutting offer very consistent prep shapes and weight. Manufacturers of these machines advertise weight consistency within 5 %.

8.3   Operator Influence and Automation

Most compression molding processes are manually operated, because it is extremely difficult to automate prep loading. TSE preps are inconsistent, sticky and flexible, making handling with automation cumbersome. Green strength plays an important role in being able to handle uncured material. Preps are usually placed into the mold cavity by hand or by use of a loading tray. In either case, an operator influences the mold-loading time.

Demolding further discourages automation. Consistently removing elastomers from a mold is difficult, regardless of the molding process chosen. However, of the available molding processes, compression molding may be the most complicated to automate. As described above, compression molding "flushes" the cavity with excess material resulting in unpredictable and unmanageable flash. The flash needs to be removed either with the molded part as a secondary operation with air blow off, or physically with a brush.

8.4   Material Flow

An operation heavily influenced by an operator requires a process robust enough to cover worst-case scenarios. This means that the process needs to produce the same good parts with a broad range of operator influence (the slowest as well as the fastest operator). Generally speaking, for this reason, compression molding processes use TSEs with substantial scorch safety.

Scorch is a term used where a TSE starts to crosslink prior to completely filling the mold cavity. Compounds designed for compression molding generally offer the most forgiving process in terms of scorch safety. With the fact that cold (room temperature) preps are loaded into the mold, cycle times can easily extending beyond what is seen in other molding processes. For the uncured TSE to reach its recommended cure temperature takes time. For instance, a hockey puck compound may have a suggested cure profile of 5 minutes at 350 °F. So the mold temperature is set at 350 °F and a room temperature (70 °F) prep is placed into the compression mold. After the mold closes and the prep conforms to the mold cavity, it may take several minutes for the center of the puck to reach 350 °F. Only then can the cure time start. Therefore, the ultimate cure time of the hockey puck can be as long as 7–8 minutes.

Conventional wisdom would suggest that a material with low viscosity is preferred over a high viscosity material. However, this is not the case with compression molding. Compression molding requires a significant amount of backpressure to pack out the cavity. If a material flows with very little resistance, it will leak out of the cavity before the cavity is completely filled. Backpressure forces the uncured TSE to remain in the cavity and develop pressure to fill corners and undercuts. Since the distance for material to flow for a compression molding operation is very short, and compression molding requires backpressure, compression molding is a preferred method for molding high viscosity materials.

Further zones influencing backpressure can also be incorporated into the mold design. This is particularly important if low viscosity materials are used for compression molding. Figure 8.5 shows a compression mold that has a backpressure zone. In this case, as the mold closes and squeezes the uncured TSE, the top insert engages with the bottom insert at the backpressure zone. Tight clearance (0.006 to 0.010″) forces the TSE to stay inboard of the backpressure zone and creates internal pressure. The uncured TSE is forced into every corner of the cavity and once the cavity is entirely filled, the excess starts to bleed through the backpressure zone.

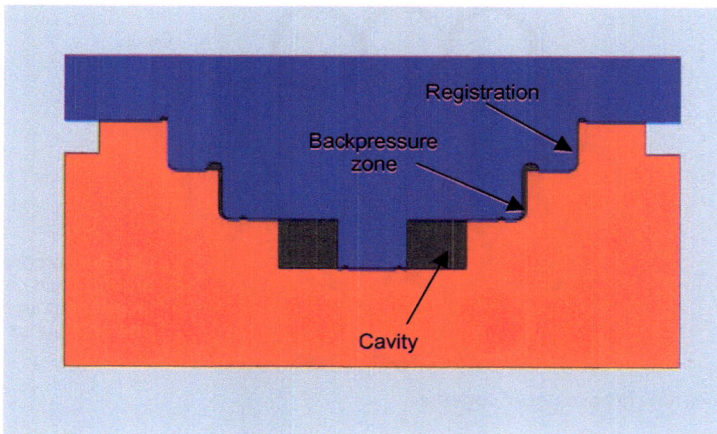

Figure 8.5 Compression mold showing backpressure zone

8.4.1   Trapped Air

Trapped air can cause problems not only with compression molding. However, unique to compression molding is the ability to minimize trapped air by altering the prep shape and weight to effectively force air to escape the mold cavity (as described above). If prep shape alterations alone do not remove trapped air, slowing the close speed of the press may allow enough time for air to escape. Together with bumps — quick opening and closing of the parting line after a closing pressure is achieved — usually takes care of trapped air. If trapped air persists, a vacuum box as described in Chapter 7 may alleviate the problem, or the mold can be redesigned to offer a parting line at the location where trapped air repeatedly occurs.

8.4.2   Molecular Orientation

Overall, compression molding can offer the most homogeneous part. Flow patterns can influence the orientation of molecules. Sprues used on transfer and injection molding create directional stresses and potential knit lines. Compression molding, on the other hand, creates few orientation stresses, resulting in a very random molecular orientation. Random orientation creates a molded article with omni-directional strength. After flow ceases in any molding process, the onslaught of cross-linking occurs. Once cross-linked, molecular orientation is petrified into the article.

8.5   Mold Construction

Compression molds can be the simplest and least expensive molds to construct. Cut-in-plate, two-plate molds are common for compression molds. Simple two-plate molds can be constructed from 4140 pre-hardened steel. Chrome plating can extend the life of the mold

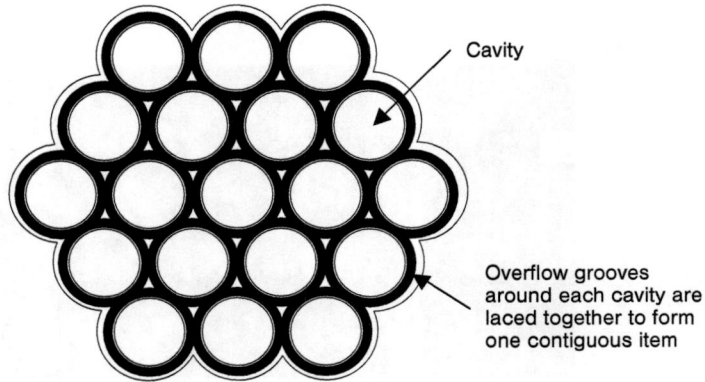

Cavity

Overflow grooves
around each cavity are
laced together to form
one contiguous item

Figure 8.6 Overflow grooves laced together

and may assist in demolding. Since most compression molds require overflow grooves, typical cut-in-plate molds can have overflow grooves that lace together (see Fig. 8.6). Not only does this allow cavities to be placed close together, but overflow groove removal is easy because all grooves are laced together, and pulling one groove removes them all.

All cavity areas, including designed (or potential) overflow surfaces, should be contained within the main hydraulic cylinder diameter. Excessive flash will increase parting line surface area and tend to keep the mold from closing entirely. Square cavity patterns where corners migrate outside of the ram diameter should be avoided.

8.5.1   Disk Springs

If cavities are inserted, high pressure disk springs are a good idea to assure that closing pressure is distributed equally among all cavities. With high pressure disk springs, the cavities can be outside of the main hydraulic cylinder. However, thicker back-up plates may be necessary to prevent mold plate bending. A few words about high pressure disk springs: these disks are heat treated and can apply high pressure under deflection. To prolong the life of disk springs, deflection should be limited to 30 %. This may not be intuitively obvious when designing a mold. The mold should not be designed to allow the inserts to bottom out on the disk spring based on system pressure; meaning, the pocket under the mold insert needs to be designed so that the disk springs are incapable of collapsing beyond 30 % deflection. Otherwise, over-loading a cavity, or forgetting to remove a molded article and remolding can over-compress the disk spring. For calculating disk spring pressure under each compression insert stack, 80 % of the total clamp force should be divided among all cavities. For instance, if 800 000 lbs. of clamp force is available for a 100 cavity mold, 6400 lbs. of force should be used under each cavity insert.

Disk springs can be stacked in parallel or series, or as a combinations of both (see Fig. 8.7). Stacking springs in parallel increases the deflection force by a factor of the amount of springs used in the same direction. Stacking springs in opposite directions, or series, increases the

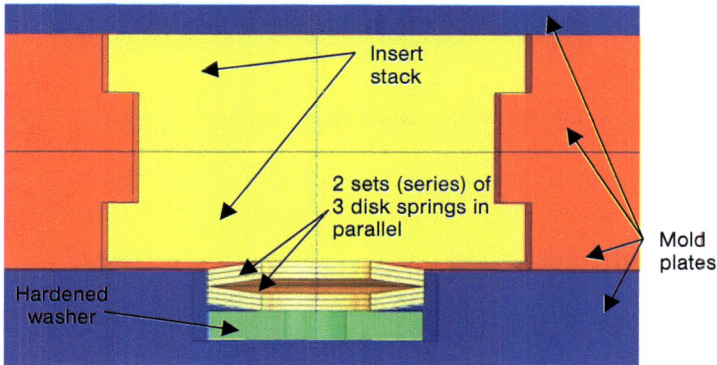

Figure 8.7 Disk spring loaded insert stack

potential deflection distance. It is important to note that if using a combination of series and parallel stacks, the parallel segments all need to include the same amount of springs, or over-deflection occurs in the segments containing fewer springs. Also, disk springs should only be used with cavity inserts that are made from hardened steel, and if pocketed in mild steel plates, hardened washers should be used to avoid hobbing into the softer steel.

8.5.2   Relative Cost

Compression molding can offer the lowest investment for entrance into production. Needing to place a prep into the mold requires a vertical platform, offering the potential for a very simple and inexpensive press.

8.6   Pressure

A good rule for compression molding is to apply roughly 2000 psi cavity pressure. To calculate cavity pressure for compression molds can be elusive. Compression molds typically flash, and if this flash is allowed to extend along the parting line parallel to the clamping parting line, the molded surface area is increased, resulting in lower cavity system pressure. Vertical dams (or backpressure zones), or large dump grooves that are designed to not completely fill, should be used in compression molds to reduce potential molded surface area. Increasing surface area perpendicular to the clamping parting line has no adverse affect on cavity pressure. Regardless, calculating cavity pressure should err on the side of too much surface area rather than too little.

$$C = B \times S$$

$$V = C/A$$

where

C = clamp force (lbs)

B = clamp cylinder area (in^2)

S = system pressure acting on clamp cylinder (psi)

V = cavity pressure (psi)

A = total horizontal molded surface area (including flash and overflow grooves (in^2)

8.7 Backrind

Backrind is a condition predominantly, but not exclusively, seen in compression molding. The symptom of backrind manifests itself in the form of a frayed edge at usually one, but sometimes all, of the mold parting lines. It is actually a thin layer of sheared material that is parallel to the mold parting line. Backrind is caused by crosslinked TSE being forced out of the mold cavity, through the mold parting line, because of thermal expansion of the TSE. It stands to reason that placing a room temperature prep into a hot mold and closing the mold will cause the prep, in time, to heat up to the mold temperature. If the TSE starts to skin cure while the inside of the prep temperature continues to rise, the prep will require a greater volume of space. TSEs do not compress, therefore the internal pressure created because the prep is expanding volumetrically, forces TSE out of the cavity at the path of least resistance — the mold parting line. Backrind is more prominent in cavities that have a high volume of space. The greater the volume of TSE in the mold cavity, the greater the chance of backrind.

Backrind can be minimized by lowering mold temperature, preheating the prep, retarding the initiation of crosslinking in the TSE, adding bumps to the molding cycle, or in severe cases, adding a pad (roughly 0.020″ thick) around the periphery of the parting line and trimming it off as a secondary operation.

8.8 Mold Cleaning

Most compression molds can withstand the rigors of even the harshest mold cleaning methods. Complicated molds with precision fit sliding inserts such as those described in the following, may be an exception. Full mold disassembly may be best for these types of molds.

8.9 Article Removal/Ejection

Article removal for many compression molds may be as simple as using an air hose to aid a gloved operator to peel articles from the mold. Articles captured in plates may need to be pushed out by means of a knock-out board. Other methods may incorporate a stripper plate, which usually has an angle fit to one half of the mold, and when lifted, strips the part from the mating plate or insert. In most cases, an extensive amount of flash comes with the parts, or is stuck to mold surfaces. Operators are tasked with brushing and removing any residual flash before loading the next cycle.

8.10   Compression Mold and Die-Cut

Although viewed as archaic and simplistic, compression molding TSE can be very efficient. For instance, in the medical industry, luer tip caps, syringe plunger seals, injection sites, and duckbill valves are often compression molded. Some of these parts see astonishing annual usages of more than 1 billion pieces.

For high volume applications, several molds stacked in a press may be required. Large 2-plate compression molds with cavities laced together with a flash pad allow for effortless part removal and very quick mold-open time. Minimizing the mold-open time reduces the risk of mold temperature fluctuations — one of the greatest contributors to molding scrap. Another benefit to this type of molding is that the TSE flushes through the cavity and forces gases and contaminates into the flash pad, resulting in very low potential scrap. To further minimize the potential for trapped gasses, presses equipped with vacuum boxes are typically used.

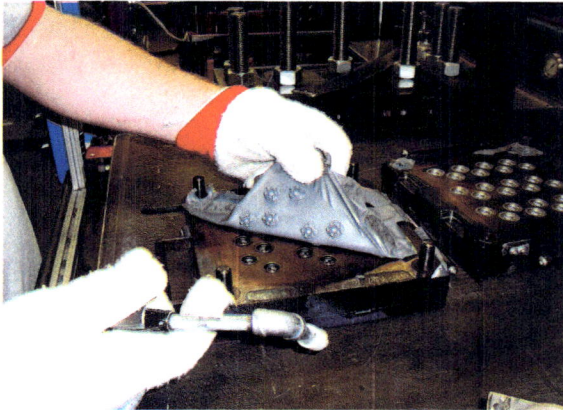

Figure 8.8 Compression molding in-sheet

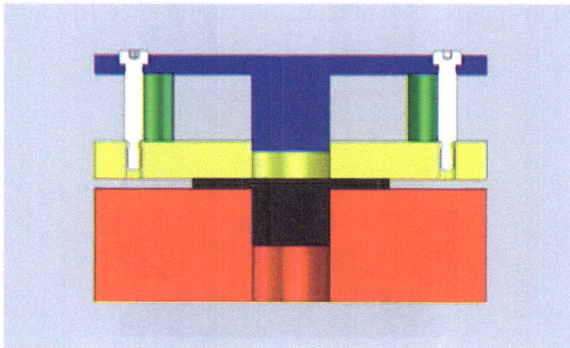

Figure 8.9 Die cutting (before cut)

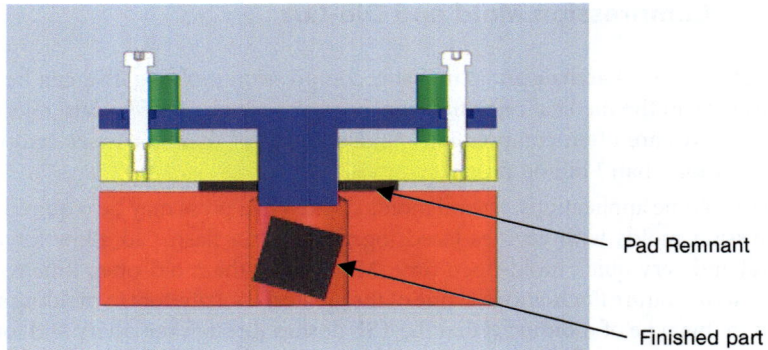

Figure 8.10 Die cutting (after cut)

Once the pads with attached parts are removed from the mold, the pad is allowed to cool, and is then placed in a punch die. This operation removes the part from the pad and can also form a valve slit simultaneously.

Compression mold and die cut parts are often used in medical applications where zero-flash is required. Since the parting line of the mold is in essence a pad, technically there is no flash, and therefore for applications where dislodged flash can create catastrophic results, die cut parts are the best answer.

8.11   Dual Acting Spring Mechanism Compression Molding

Unlike common compression molding, dual acting spring mechanism (DASM) compression molds use heavy spring-force to hold mold cavity inserts closed before TSE fills the cavity. As the mold closes, high pressure (disk) springs act on a sliding mold component, which in turn coin on a molding insert, or mold shut-off (see Fig. 8.11). The exerted spring force is calculated to overcome the potential cavity pressure and is said to be "flashless." Total spring force should be held to approx. 75 % of total press clamp force. In addition, the travel of the

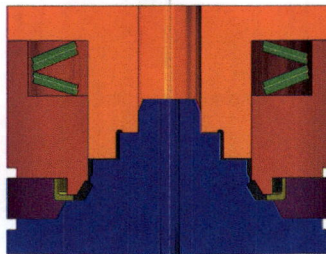

Figure 8.11 DASM compression stack

outer sliding insert needs to be calculated based on the prep height to allow for adequate pressure prior to filling the cavity (usually about 0.15"). The outer sliding insert does not necessarily have to make contact before the inner contacts the TSE prep; it just needs to establish contact and adequate pressure prior to the cavity fill. The initial closure can act on flashless grinds and offer the same advantages as flashless transfer molding. The final closure of the mold acts on the uncured TSE and forces the TSE into the cavity. This process can be very efficient. In fact, many automotive shaft seals are manufactured using this method.

Figure 8.12 shows the mold-closing in a DASM compression mold. A precision donut-shaped prep is placed into a pocket in the mold. The pocket has precise clearance on either side to create a dam, or backpressure zone. This "packs" the TSE into the cavity. The inner piston slides inside of the outer insert every cycle. Vertically moving inserts that are in contact with the pressurized TSE, and cannot be brushed/cleaned each cycle, require extremely tight clearances (usually 0.0002/0.0006"). These moving parting lines require precision ground surfaces to prevent gauling. A Teflon seal is used to prevent TSE from flowing between the mold inserts and offers added friction resistance to prevent binding. It stands to reason that DASM designs require hardened tool steel. Typically, all mold inserts are chrome plated and are constructed from 440 SS hardened to 46–48 Rc.

The donut shaped prep — or I. D./O. D. prep — needs to display consistency in weight and shape. Prep height and wall thickness throughout the prep is crucial to successful molding. Preps with thinner cross-sections will generally non-fill in that same location in the mold. Preps are usually prepared on a pre-form machine as described previously. In the case of DASM preps, a die with an inner and outer section (sometimes referred to as a spider die), extrudes a tube that is cut with a sharp knife to the desired weight/thickness.

Undercuts can be added to the mold at the vertical fill section of the top outer insert to pick the gate remnant from the molded part as the mold opens (see Figs. 8.13 and 8.14). The vertical section of the gate is locked in between the top and bottom insert. As the mold opens, if an undercut exists in the top insert, the force will tear the gate clean. For most materials,

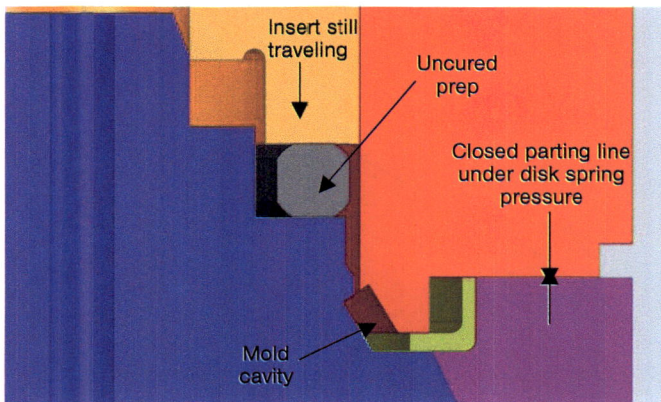

Figure 8.12 Detail of DASM closure

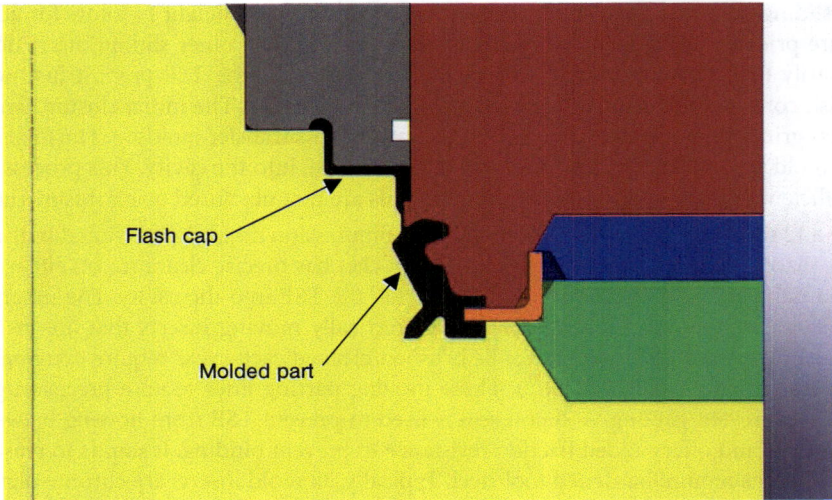

Figure 8.13 Detail of self-trim

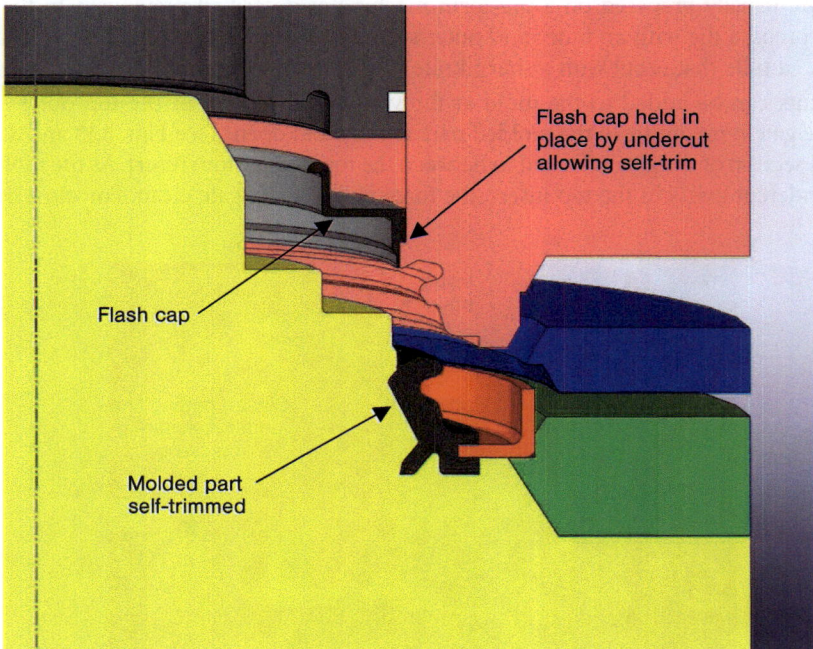

Figure 8.14 Detail of self-trim open

the gate needs to be held to approximately 0.004″ (dependent on TSE being used) clearance for a clean break to occur.

8.11.1   Prep Compensating Mechanism

The DASM mold design described allows excess material to flow through the backpressure zone and ultimately into dump grooves. Prep weight variation is simply compensated by allowing excess area for heavier preps to fill.

Prep weight compensation can also be achieved by incorporating disk springs behind the secondary sliding insert (see Fig. 8.15). The mold should be designed to allow the disk springs behind this sliding insert to maintain 2 000–3 000 psi cavity pressure. The travel of the prep compensation insert should be calculated to allow for adequate cavity pressure throughout the prep weight tolerance (usually about 0.060″ total travel). This method allows for a more

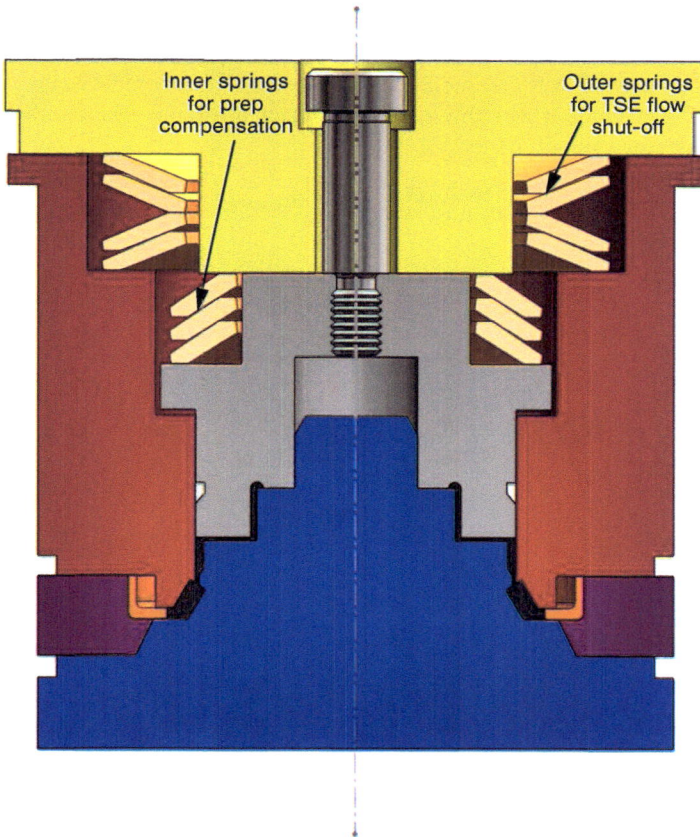

Inner springs
for prep
compensation

Outer springs
for TSE flow
shut-off

Figure 8.15 Detail of DASM with prep compensation

consistent cavity pressure, as opposed to simply allowing material to fill an overflow groove. Also, as the TSE material temperature changes, its volume increases. The prep compensation insert will conform to this thermal expansion without dramatically altering the cavity pressure. As with all DASM designs (single or double spring packs), total spring force should be held to approximately 75 % total press clamp force.

8.11.2   Secondary Trim

DASM molds can also be designed to have the gate remain with the molded part instead of breaking free during the mold opening. In designs where the part picks clean, the gate remnant is left in the mold and needs to be removed. An air blast can usually remove all excess TSE, but slows down the molding cycle and cools down the mold. If the gate remains with the molded part, the demolding cycle can be reduced, and potential contamination avoided.

With many materials, shear stresses incurred from a 0.005–0.007″ gate require a cooler running mold to avoid scorch. If sheer stresses can be reduced significantly, for instance by a 0.015–0.025″ gate, mold temperatures may potentially be increased substantially. Therefore, larger gates may allow hotter molds, which may significantly reduce cure time. However, a 0.015–0.025″ gate will not self-trim in a mold. In fact, it will not even hand-tear trim easily, even with poor tear strength materials.

An advantage to thicker gates is quicker cure times, but at the expense of a secondary trim operation. Knife trimming is a common secondary operation for many shaft seals. Some shaft

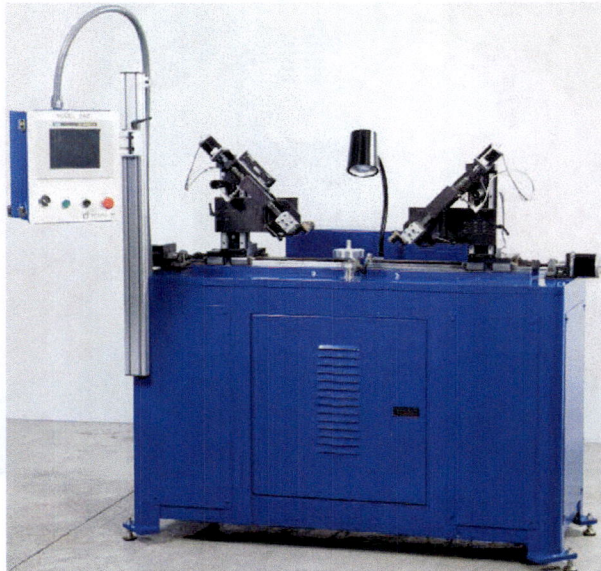

Figure 8.16 Shaft seal trimming machine (Courtesy of Mitchell, Inc.) [3]

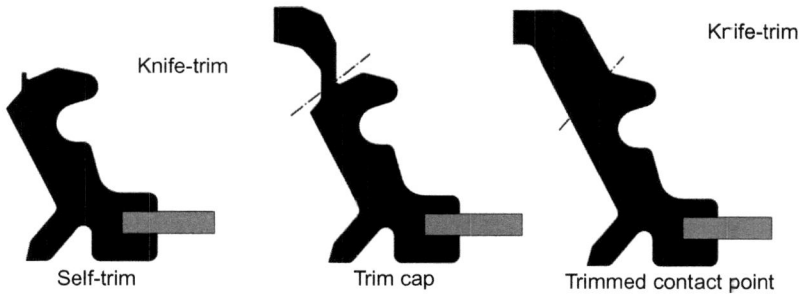

Figure 8.17 Shaft seal trimming

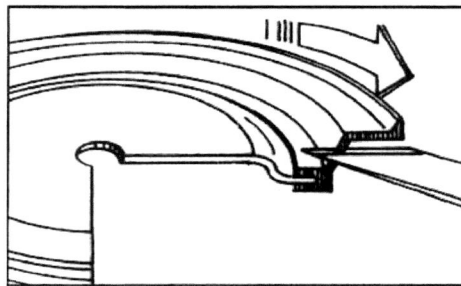

Figure 8.18 Trimming detail view (Courtesy of Mitchell, Inc.) [3]

seal companies prefer to knife trim the contact point, and claim the sharp contact point adds to sealability in the application. In most applications where the lip is trimmed, there really is no gate, so sheer stresses can be greatly reduced.

Knife trimming is usually performed with an automated, or semi-automated machine (see Fig. 8.16) that utilizes a nest to house the molded seal. The nest spins with the shaft seal in place and a carbide knife penetrates the gate and removes the remnant. TSEs are abrasive, and therefore knife design and sharpness are paramount, if knife trimming is the preferred method, particularly if the contact point of a shaft seal is to be trimmed. Phonograph grooves on the trimmed surface of a shaft seal can act as a pump when the shaft is spinning and can either run the contact point dry, or pump oil past the contact point and cause a leak. A strict knife replacement/sharpening plan should be administered.

8.12   Conclusion

As depicted, compression molding spans in complexity from low-volume, simple, low-cost tooling, all the way to high-volume, complex, high-cost tooling. It simply is what the custom molder wants it to be. Compression molding could be considered the most versatile process and certainly a competency that every custom molder should have to offer diversity in methods and volumes.

References

1. The French Oil Mill Machinery Co., Piqua, OH.
2. Barwell Global, LTD, Cabs. U. K.
3. Mitchell Inc., Elkhart, IN.

9   Transfer Molding

Transfer molding was the first closed-mold fill method for TSEs; whereas compression molding, the first molding method, was considered an open-mold method. This chapter will cover self-contained transfer, bottomless pot transfer, flashless transfer, and wasteless transfer molding methods. In addition, process related topics will be covered. Transfer molding is still widely used in custom TSE molding. The methods discussed, in some cases, can offer more efficient and cost-competitive processes than injection molding. Some techniques can be used in conjunction with injection molding that take advantage of both methods.

Common to all transfer molding techniques is the use of a pot and piston (plunger). Uncured material is placed in the pot and the press closure (or auxiliary cylinder) forces the material into the cavities.

9.1   Self-Contained Pot

The simplest form of a pot and piston is accomplished by machining a "pot" or pocket into the top "cavity" plate of a mold (see Fig. 9.2), and machining a mating piston into the top "mounting" plate. Inside of the pot, small holes are drilled (sprues) which connect to designated locations in the cavity. This is commonly referred to as self-contained pot transfer molding.

In self-contained pot transfer molding, uncured TSE preps are laid in the pot. The piston enters the pot as the press closes, and forces the TSE to close the mold cavity parting line. As the pot continues to close, the TSE is forced through the sprues and fills the cavity. Heavy die springs are included between the pot and piston plates to assist in opening. The self-contained pot is roughly 0.50″ deep, and the piston is approx. 0.47″ high, leaving a 0.03″ gap when the mold closes. Deeper pot and piston combinations increase the chance of binding. The gap created by the shorter piston will result in a pad of cured TSE (called pot scrap) with attached

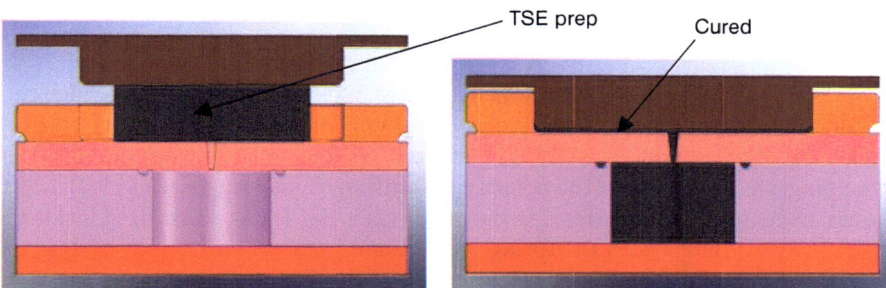

Figure 9.1a Transfer mold closing   **Figure 9.1b** Transfer mold closed

Figure 9.2 Transfer mold in press

sprue remnants after completion of a molding cycle. Since the pad is discarded, it is tempting to reduce the gap between the pot depth and piston to use less material. This however, could result in the pad tearing or sprues breaking off and staying in the mold. Removing the torn pieces delays the change time, so careful consideration should be taken to create a pad that is not too thin.

One way to obtain thinner pot scrap without losing pad integrity is to use paper sheets. The paper sheets are placed over the uncured preps. The TSE will cure onto the paper, offering increased integrity.

Self-contained pots can have a variety of patterns; square and round patterns are most common. Self-contained pots have a tendency to bind, gall, and leak. Patterns other than round only accentuate this condition. Therefore, corners on non-round pots should have generous radii. Pistons are often bronzed to offer a soft half of the mating fit to reduce galling. As the piston wears, paper (as mentioned earlier) can be used to seal the clearance between the pot and piston.

9.2   Bottomless Pot Transfer

The bottomless pot transfer molding method has a large, typically round, spring-loaded ring called a pot, which hangs from a precision mated piston mounted to the top platen (see Fig. 9.3). This ring has no bottom plate, and the process is therefore termed bottomless pot transfer molding. On the bottom of the pot is an inwardly machined ring (see Fig. 9.4), which generally has an angled bottom surface. As the press closes, the TSE is pressurized inside the pot, and TSE pressure along with the spring pressure forces the inward ring to seal on the top plate of the mold.

As the press continues to close, and more pressure builds inside the pot, the inner piston insert is allowed to slightly push back into the main piston. This insert has an angled outer edge which mates with another ring of softer material. As the inner seal is pushed up, the outer

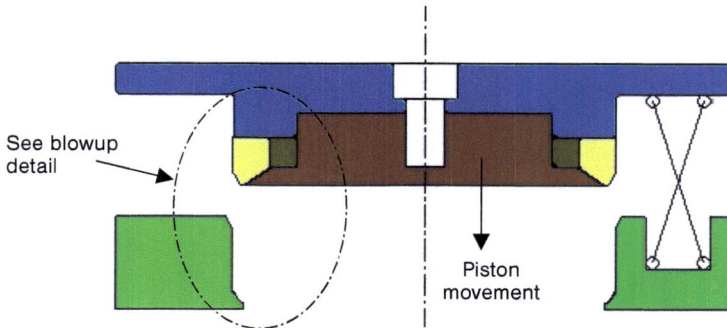

Figure 9.3 Pot and piston shown in open position

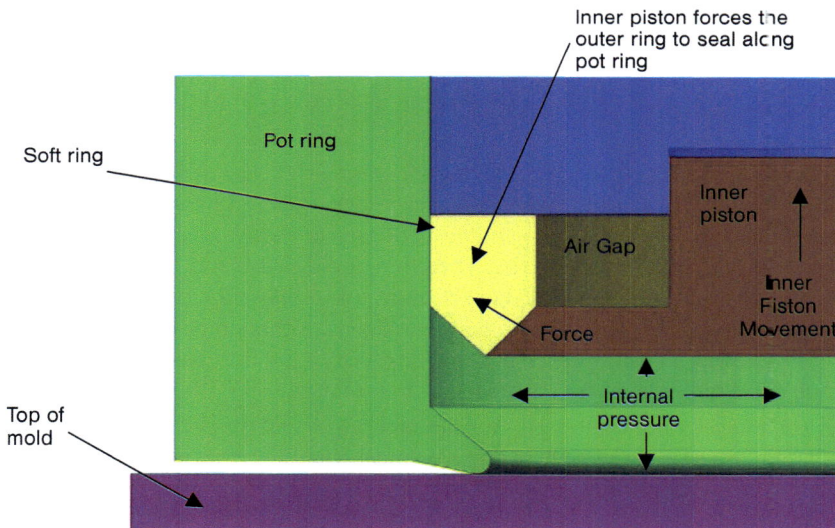

Figure 9.4 Pot and piston detail

angled surface exerts pressure on the softer ring and forces the ring outward to seal tightly against the pot ring. This sealing system avoids TSE leakage between the pot and piston. The outer ring should be frequently monitored and replaced when wear is detected.

Bottomless pot transfer molding offers a versatile arrangement. The pot and piston usually remain in the press while a number of molds can run interchangeably under the same pot. Mold changes are minimal as typical transfer molds are simply mounted to a push-pull cylinder, or manually pulled in and out of the press. Molds can be preheated on auxiliary platens allowing for mold changes in just a few minutes.

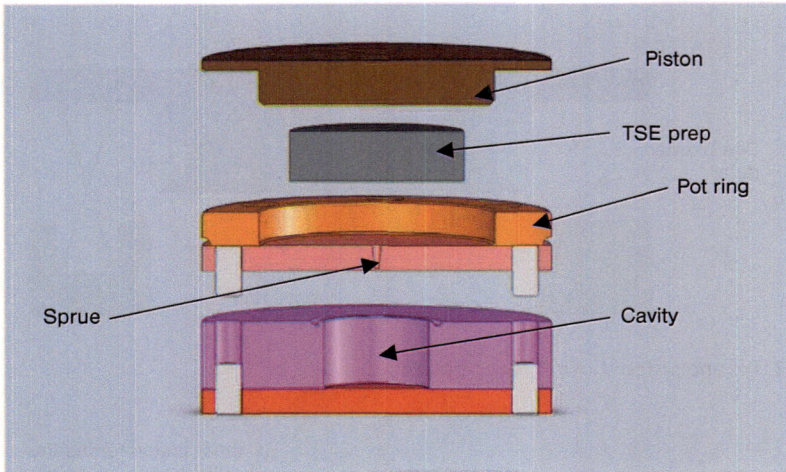

Figure 9.5 Transfer mold exploded view

Bottomless pot transfer molding can be very forgiving in terms of parallelism of the mold frame and press, particularly if individual inserts are used and allowed to float within the mold frame. With individual inserts, internal pot pressure activates each cavity individually. The mold frame only acts as a carrier. Therefore, mold plates and platens can dish in the process, yet the quality of the molded article may not be affected. Using a bottomless pot in the same press long-term can seriously crown platens. Although not as critical for transfer molding, it is unwise to use a press that has a long history in bottomless pot transfer molding for compression molding where parallelism is paramount.

Another advantage of bottomless pot transfer molding is that the pot scrap is easily removed since it lies on top of the mold where it can be quickly removed as one intact piece. Material waste, on the other hand, is a shortcoming of this process.

9.3   Transfer Press

Transfer molding can also be done with a pot and piston incorporated into the press. In this instance, a hole is drilled through the bolster and top platen of the press. A cylinder is attached to the top bolster plate of the press and a piston attached to the cylinder has a precision fit to the hole in the platen (see Fig. 9.6). Uncured TSE is placed on the top surface of the mold. The mold has a runner and sprue holes in the top plate which lead to the mold cavity. The press is closed to a set clamp pressure, and then the cylinder is activated and the piston squeezes the material into the mold.

Figure 9.6 Transfer press (Courtesy of Fullwell Holdings) [1]

9.4   Flashless Transfer Molding

This technology lends itself well to round parts that can be either all-TSE or include bonded carriers. Each cavity is machined individually and inserted into holding plates. Each cavity can be considered as an independent mold — meaning, the action of one cavity does not influence the activity in an adjacent cavity. The top insert of the cavity stack has a slip fit (0.000 4″) with a bore in the top holding plate. A groove is machined diametrically near the top (pot side). Additional clearance is machined above the groove, leaving a small gap

Figure 9.7 Flashless transfer cavity set (Courtesy of Fort Wayne Mold and Engineering, Inc.) [2]

between the bore of the top plate and the top insert, which allows TSE to fill the groove when the first load is run (see Figs. 9.8 and 9.9). The TSE that fills the groove remains in place until the mold is disassembled. The molded ring is commonly referred to as a rubber lock. The rubber lock holds the top insert in place, yet allows it to move slightly up and down under pressure. The rubber lock also creates a concentration point for the rubber to break instead of allowing flash to migrate between the insert and plate.

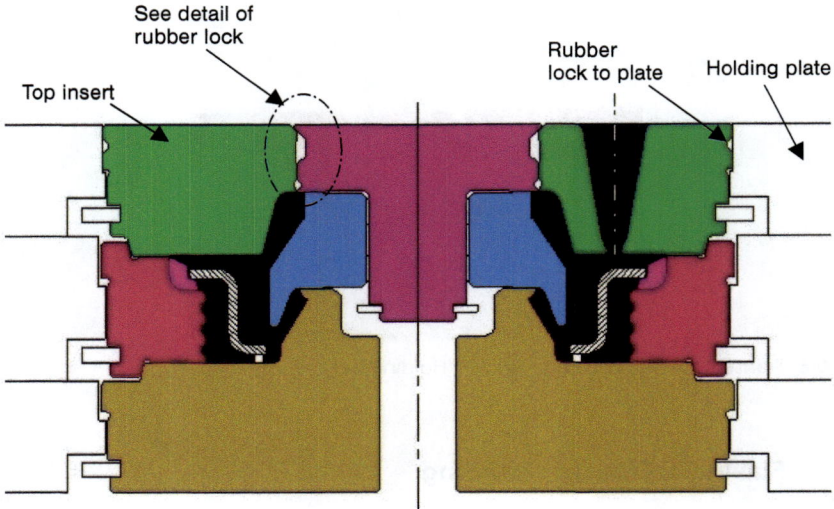

Figure 9.8 Typical flashless transfer cavity stack

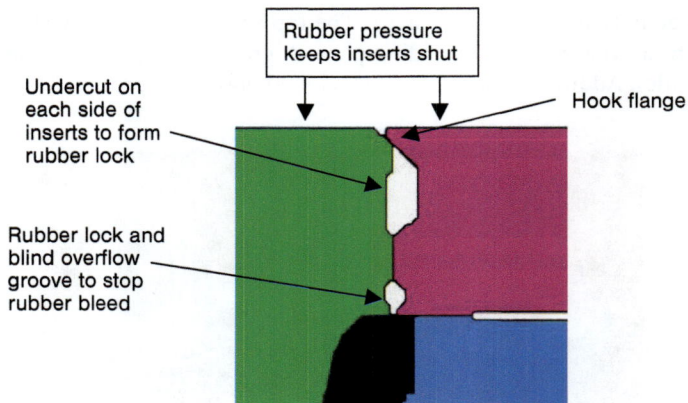

Figure 9.9 Detail of flashless rubber lock

Other than the top insert, the remaining inserts have plenty of clearance in their respective holding plates, allowing them to float freely. Each insert has a vertical or angular registry to the next mating insert in the cavity stack. Therefore, each cavity stack is considered self-registered. The cavity stacks have a taller profile than the plate stack up. This assures that the inserts column out instead of the plates.

Each horizontal parting line on the cavity stack has one mating surface ground parallel and smooth, while the other side has a rougher grind (either parallel or at $-\frac{1}{2}°$ taper), usually 28–32 micro-inch finish. The top plate of a transfer mold has holes bored through for housing the top insert. When assembled, the top of the top insert is exposed. When TSE is introduced into the mold, uncured preps are simply placed on the top of the mold. Typically a bottomless pot (as described previously) is utilized in conjunction with this mold design.

Flashless transfer molding can be understood through the principles of hydraulics. As mentioned in previous chapters, before the TSE crosslinks and heat is applied to the process, the material can be considered fluid-like. Envision the pot and piston above the mold as a hydraulic cylinder. Pressure is exerted in every direction inside the pot. Therefore, each cavity stack is subjected to the identical pressure as its neighboring cavity. More specifically, the top insert of each cavity stack is subjected to the system pressure inside the pot. Recall the construction of the mold pertaining to the top insert and its relationship to the top plate. As pressure is applied to the top surface of the top insert, it is allowed to slip vertically in the bore of the top plate. The top insert diameter must be calculated to adequately account for the internal cavity area. This calculation is necessary in order not to allow the internal cavity pressure to overcome the pressure applied to the top surface of the top insert and consequently blow the mold insert's parting line open. Each cavity stack can be interpreted as having its own hydraulic piston closing the cavity stack. In this case, it is the top insert which is acting as the hydraulic cylinder, exerting a force on all of the parting lines of the cavity stack.

The force exerted on the top insert is transferred to each parting line. When the rough ground parting line surface is forced against the smooth ground parting line surface, the parting line is considered shut — at least as far as the TSE is concerned. The microscopic grooves in the rough surface are ground to a specific depth to allow air to escape, yet prevent TSE flow. Therefore, TSE flow across the parting line is almost non-existent. Herein lies the premise of flashless molding.

9.4.1   Split-Top Inserts

In cases where the molded article has no center (hollow/donut shaped) as in Fig. 9.8, a split-top insert is required. This allows hydraulic pressure to act independently on the I.D. and O.D. cavity configuration. Since it is nearly impossible to time the inserts to have perfectly identical stack-up's, splitting the top insert enables the internal pressure of the pot to compensate individually on the inserts. The split-top insert also requires a rubber lock as used with the top insert-to-plate interface described earlier.

With a split-top insert design, the center insert is only hanging in place by the rubber lock. In cases where the center insert has a cavity geometry that may require to overcome high forces during the opening cycle of the mold, and/or a hanging insert (as shown in Fig. 9.8), a hook flange should be used to assure that the center insert does not slip through the outer insert (a hook flange is shown in Fig. 9.9). Under the hook flange adequate clearance still needs to be included to make sure TSE can flow to the rubber lock and that the hook on the inner insert does not bottom out on the outer insert.

9.4.2   Vents

Flashless vents are usually prepared in a spiral pattern where a narrow grinding wheel is directed toward the center of the insert and then moved back and forth as the insert is slowly turned diametrically. The pattern in Fig. 9.10a is a result of such a vent. A more dramatic vent can be accomplished by grinding along the tangent edge of the cavity. In this case, the grinding wheel is directed back and forth along the tangent cavity edge as the insert is turned diametrically. This is commonly referred to as a tangential grind (see Fig. 9.10b). Recently, laser venting has become popular. Laser vents can be specifically located as opposed to venting

Figure 9.10a Ground flashless vent

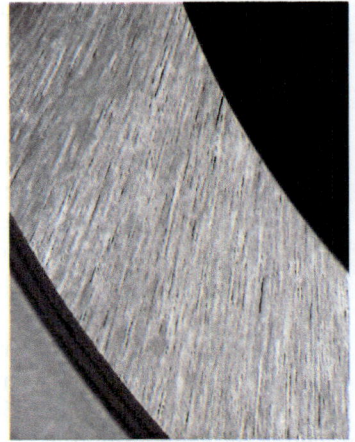

Figure 9.10b Tangential ground flashless vent

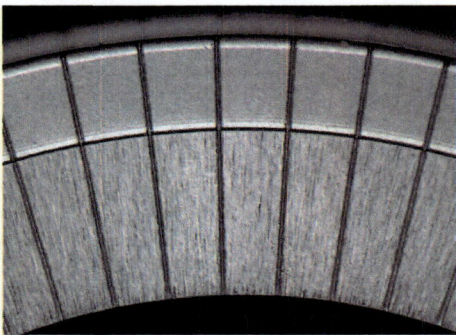

Figure 9.10c Laser flashless vent

(Courtesy of Fort Wayne Mold and Engineering, Inc.) [2]

an entire surface with grinding. Also, the depth of the vent can be varied along the length of the vent. Laser vents also have a more accurate vent depth. Figure 9.10c shows a typical laser vent. Depending on the TSE to be molded, vent depths vary from 0.0001 to 0.000 5″. The tolerance of the depth can be held to ±0.000 05″.

9.4.3   Trapped Air

Trapped air is notorious in flashless transfer molding. The whole premise of flashless transfer molding is to hold parting lines shut. Air is allowed to vent through the microscopic grooves, while TSE is held at bay. There are many parting lines on typical flashless molds to allow plenty of air to escape. Vacuum boxes are highly recommended to minimize trapped air. Many applications can reduce trapped air by slowing the close speed to allow enough time for air to escape. Bumps are generally not recommended in flashless transfer molding, since clamp pressure loss can lead to flash. However, in severe cases, pressure bumps may be effective. Unlike conventional bumps, pressure bumps only relieve a percentage of clamp pressure, never completely losing clamp pressure, thus preventing the TSE from leaking across parting lines.

9.4.4   Sprues/Gates

Usually sprues can easily be added to help fill areas starved for material; particularly in conventional transfer molds, where the pot floods the entire top surface of the mold. In essence, the pot is a large runner system, and adding sprues requires no modification to this runner system.

Sprues should be designed with a short straight (non-tapered) hole at the entrance to the cavity (see Fig. 9.11a). This allows for an accurate/consistent diameter. Also, as the sprue wears, in the absence of a straight, the sprue diameter will increase by a factor of the taper. Sprues should have substance so that during the removal process, they will not break and leave a portion of the sprue in the sprue hole. Sharp corners at the transition to the pot should be rounded to eliminate stress risers.

If sprue remnants left on the molded article are a concern (high sprue), a dimple or boss can be included at the sprue location to keep the remnant sprue height below the surface of the molded article (see Fig. 9.11b).

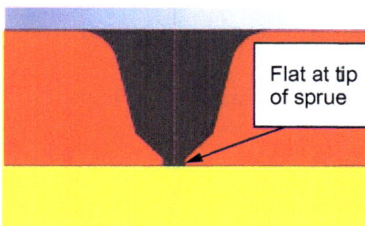

Figure 9.11a Standard sprue

Flat at tip of sprue

Figure 9.11b Recessed sprue

Raised pad in mold leaves recess in part

Figure 9.12 Dummy sprue

Sprues can be used not only for filling cavities, but to pick cavity edges clean (without flash residual). For instance, parts that have small holes can be designed to include pins in the bottom half of the mold and a sprue positioned directly above the pin (see Fig. 9.12). The sprue diameter needs to be slightly smaller than the pin; otherwise the sprue remnant can pick into the molded article. The pin can be ground shorter than the mold closure, offering clearance for the TSE to fill the cavity — in essence forming a gate. Alternatively, the pin can contact the top insert, offering no clearance. This is commonly referred to as a dummy sprue. In either event, when the sprue is removed, the sprue picks the hole clean, leaving no flash extension.

The dummy sprue concept can extend beyond just a round hole. Creative machining can include a gate that follows the contour of the molded article. The result will allow for a flash free contour once the gate is removed.

9.4.5   Knit Lines

Most transfer molded articles use sprues to fill the cavity. Multiple sprues are often used to limit the distance the TSE is required to flow. As flow fronts come together during the cavity filling cycle, knit lines can result. The flow fronts can contain contamination picked up in the mold, such as mold release, or the flow fronts can simply exhibit a skin cure. In either event, an ineffective welding of the fronts may occur. Knit lines are not always considered a defect. Depending on the end use of the product, the anomaly may be considered anything from a visual blemish to a catastrophic defect.

Knit lines can be minimized by lowering the speed of cure and/or minimizing flow restrictions. Cure-related remedies include lower mold temperatures, or compound cure system revisions. Flow-related fixes include slower speeds, larger (or more) sprues, or lower viscosity compounds.

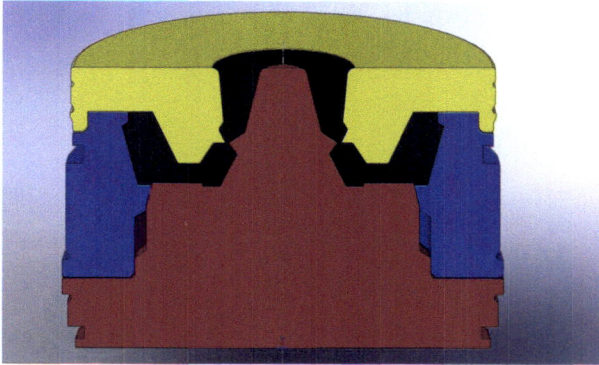

Figure 9.13 Ring gate insert stack

9.4.6   Ring Gates

Ring gates are often used where sprue vestiges and/or knit lines may not be allowed on the molded article (see Fig. 9.13). Additionally, ring gates generally introduce less directional orientations and a more even fill. Ring gates are normally restricted to round parts. They require tight registration between mating inserts; otherwise the gate thickness may vary diametrically. Ring gate thickness varies, depending on the material to be molded. The gate should be thick enough to adequately fill the cavity, and thin enough to break clean. Cavity inserts with ring gates are more prone to damage and may require more frequent maintenance.

9.4.7   Mold Construction

As mentioned previously, most flashless transfer molds incorporate many parting lines to allow air to escape, thus requiring many mold plates. The holding plates are usually constructed with 4140 pre-hard steel. Flashless molds require a gentle cleaning process to avoid damage to the delicate grinds, so soaking entire molds in cleaning solutions is common. Mold plates are commonly flash-chrome plated to avoid corrosion.

Individual cavity inserts are typically machined from 440 stainless steel, hardened to 54–56 Rc, and finally hard chrome plated. Flashless transfer molding requires mold steel to be in the mid to upper 50's range for hardness (Rc) or vents and lands will wear out prematurely. Since many TSEs exhibit corrosive behavior and cleaning methods require soaking in solutions, stainless steel is used in case the chrome chips or flakes, exposing the bare mold steel.

9.4.8   Transfer Pressure

Internal cavity pressure for flashless transfer molding should be limited to 3000 psi, unless extremely high viscosity compounds are used. Greater pressure could destroy the flashless grinds. Internal pot pressure (which is the same as cavity pressure for transfer molding) is calculated using the following formulas:

$$C = B \times S$$

$$P = C/R$$

where
C = clamp force (lbs)
B = clamp cylinder area (in^2)
S = system pressure acting on clamp cylinder (psi)
P = pot pressure (psi)
R = piston (inside of pot) surface area (in^2)

9.5   Mold Cleaning

Flashless molds generally require more frequent mold cleaning than other molding methods because the grind/vents get plugged. Mold cleaning methods should be limited to non-abrasive methods. Therefore, air pressure with media should be avoided, even if the media is soft. The high pressure will prematurely wear the mold shut-off surfaces. Media is also difficult to remove, and if left on parting lines will damage the delicate vents. Solvent tanks with ultrasonics are the best alternative. In some cases, additional cleaners along with brushing (with soft bristles) is required to remove stubborn build-up. Molds should be rinsed with clean water as a last step.

Need for Spare Inserts

Flashless transfer molding requires a great deal of maintenance to remain flashless. For this reason, a healthy supply of spare inserts should be on hand for a quick exchange, if a cavity gets damaged or starts to flash. Flashless inserts can be reground several times before too much material is removed. Since material is removed during the regrinding of the vents, the need to requalify reworked cavities should be examined.

As soon as a cavity begins to flash, it should be exchanged. This usually means the cavity needs to be replaced at the press. Ideally, all cavity inserts should have access to retaining rings from the parting line, or allow for insert removal without unbolting back-up plates. Experienced tool technicians can replace cavity stacks in a matter of minutes, if the mold is designed properly.

Plating

Hard chrome plating is recommended on hardened 440 stainless steel inserts. Chrome lengthens the life of flashless grinds and does not hinder the venting. Other coatings can be used, but caution should be exercised with softer platings as they may hob and restrict vents.

Brushing

Flashless vents should be physically brushed at each demolding cycle. This can be accomplished by use of a hand brush or a low RPM rotary brushing machine (similar to automobile buffing). In either case, soft bristles need to be used to avoid damaging the flashless vents. Tampico, or soft plastic bristles are recommended. Flash and/or adhesive build left on the parting lines will plug vents, hob into the grinds, and perpetuate flash. Any debris left on the mold parting line will offer a fissure to allow TSE to flow through in a subsequent cycle.

9.6   Wasteless Transfer

In wasteless transfer molding, the pot is insulated from the mold. In other words, the pot is maintained at a temperature below the compound's chemical crosslinking threshold. For instance, if the mold temperature is 370 °F, the temperature of the insulated pot may only be 200 °F. The term wasteless is somewhat deceiving; although the entire pot flash is not wasted, sprues are still discarded along with a reversed sprue. The reverse sprue is the cured portion of a sprue that enters the cold pot and breaks off at the transition zone where the TSE does not cure (see Fig. 9.14). Usually sprues are captured through a fine mesh material (typically woven polyester) and pulled out of the mold in one large sheet allowing for quick removal. Some product applications restrict the use of fabric material, as fibers can enter the cavity and remain in the molded article. A metal plate with holes can also be used to capture the sprues. In this case, a knock out board is used to remove the sprues. Advantages are clear for this process. Material waste is greatly reduced, distinctly advantageous in high material cost TSEs.

Recall the design of a flashless transfer cavity insert stack as described previously in flashless transfer molding. The top insert has a tight fit diametrically with the holding plate, yet is allowed to float up and down. To retain the top insert in the plate, a rubber lock is used. For wasteless transfer molding, a rubber lock can be used as well, but will require the first cycle to be run as a conventional transfer system, where TSE will be able to bleed down into the rubber lock groove. For better retention to the holding plate, high modulus-high tear strength materials can be run as the first load. Careful consideration to cross contamination should be exercised, if different materials are used for the rubber lock. Alternatively, a shoulder and an opposing retaining ring can be used to retain the insert into the top plate. Enough clearance should be designed to allow the insert to float without bottoming out on the shoulder or retaining ring,

Customarily, several cycles of uncured TSE are placed in the pot. Therefore, several cycles can run before more material is added. Presses that run wasteless transfer molds should be equipped to adjust automatically to new closure heights as material is consumed in the pot.

A word of caution about wasteless transfer molding: uncured TSE in the pot can be a mess if it is not held consistently to the piston as the mold is open. Surface finishes, undercuts, ribs, and grooves can all be used to make the uncured TSE stick to the piston. If the uncured material sticks to the insulator board, or worse yet, a portion to the piston and the rest to the insulator board, the uncured material left on the insulator board needs to be scraped away from the

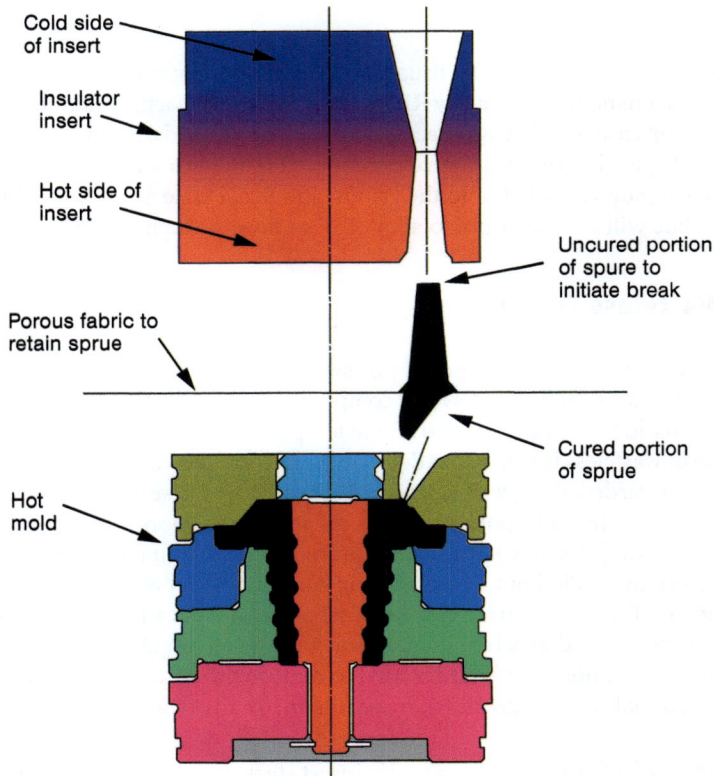

Figure 9.14 Wasteless transfer mold cavity stack

edge where the bottomless pot seals, or leakage can occur. The cycle will be lengthened by the time it takes to scrape the material, and it adds potential contamination from increased handling. Therefore, full effort to keep the TSE on the piston should be administered.

Some materials result in dead spots when wasteless transfer molded. In these cases, uncured TSE needs to be periodically scraped from the piston in the dead zones and folded into areas of the pot that offer increased flow. Varying prep placement may move dead spots and avoid the need to scrape the piston. With materials that have short residence time limits, such as materials prone to reversion, materials may need to be scraped and discarded from the piston periodically and replaced with fresh material.

One of the disadvantages of wasteless transfer molding is the absence of heat in the upper cavity portion of the mold (referring again to a vertical platform press). Adding heater cartridges to the top mold plate comes with penalties; the top mold plate usually needs to be made thicker resulting in a longer sprue, and cavity spacing is reduced because heater cartridges need to be positioned through the top mold plate spreading the spacing. Therefore, with the absence of heat in the upper portion of the mold, all heat needs to be transferred from the bottom platen. This results in mold temperature variations and longer cure times because the top plate will require a longer period of time to reach its cure temperature as opposed to having direct contact with a heat source. Additionally, wasteless transfer molding requires longer set-up times, more maintenance, and higher equipment/tooling cost. Wasteless transfer molding is generally performed on high-volume applications with expensive compounds. As with all TSE molding processes, molds should run uninterrupted through breaks, shift changes, and lunches to avoid shut-down and start-up. With wasteless transfer molding ,where temperature stabilization is critical, shutting down a mold should be limited to end-of-run, maintenance, or an out-of-control process.

9.6.1   Equalizing Platen

As described above, one of the inherent problems associated with wasteless transfer molding is the heat loss to the top cavity plate. This condition usually defaults wasteless transfer molding to a longer cure time as opposed to conventional transfer molding. While wasteless transfer molding utilizes less material, in some cases the lengthier cycle time offsets the advantages.

Another drawback of the wasteless process is that the sprue is required to fracture at the transition point between cured and uncured TSE. The sprue is positioned in the insulation zone of the mold, meaning it is located in an insulation plate housed between a heated plate and cooled plate. Potentially, cured TSE will remain in the insulation board (or insert), or continue curing during lengthy change times and get transferred into the mold in the next cycle.

In some cases, success has been achieved in shuttling an equalizing platen into the press mid-way through the cure cycle. The idea is to transfer the material through the insulator board and into the mold. After the article is "skin-cured," the cure cycle is interrupted, the press opened, and a platen is shuttled into the press between the top plate of the mold and the insulator board. The press is then closed and the cure cycle commences. The bottom half of the platen is heated, so that heat can be transferred directly to the top plate of the mold. The top half of the platen is an insulator board, and has no effect on the uncured TSE on the pot side of the platen.

Equalizing platens do come with complications. Uncured TSE can build up on the top side of the platen and may require periodic scraping and cleaning. Also, any time that clamp pressure is released during the cure cycle, flash is probable. Side cylinders and spring loaded inserts may be required to avoid pressure release and flash.

9.7   Conclusion

Transfer molding can be a very efficient molding method that offers flash-free articles, high cavitation, a forgiving process, and potentially low material waste. Transfer molding should be a capability included in a custom molders repertoire. Having the capability of compression, transfer and injection processes offers the custom molder versatility to produce a variety of products, with a multitude of materials, and at various volumes.

References

1. Fullwell International Holdings, LTD. Shatin, N. T., Hong Kong.
2. Fort Wayne Mold and Engineering, Inc. Fort Wayne, IN.

10   Injection Molding

Injection molding of TSEs can provide the most accurate and controlled method of molding. Moreover, since material is delivered into the mold mechanically, automaticn is generally easier to accomplish with injection molding as compared to other methods. Injection molding machines are configured in both vertical and horizontal platforms. Generally speaking, the injection units of TSE injection molding machines perform similarly to plastic injection machines. An injection mold will have a series of runners and sprues that lead into the mold cavity. The mastication in the injection screw, and the stresses from being forced through a runner and sprues at high velocity, preheats and lowers the viscosity of the TSE, thus offering flow and cure time advantages over conventional molding methods. This chapter describes injection molding of TSEs along with associated tooling and process considerations.

Material changes and set-up times place injection molding at a distinct disadvantage compared to other molding methods. Injection molding is usually reserved for high-volume products, or if a common material is used in a variety of molds. Dedicated molds, or a family of molds, are common for injection molding.

Injection molding can include hybrids with other molding methods, such as injection-compression and injection transfer molding. Insulated delivery methods such as cold runners can minimize waste. Injection molding is an important process for custom molders to offer.

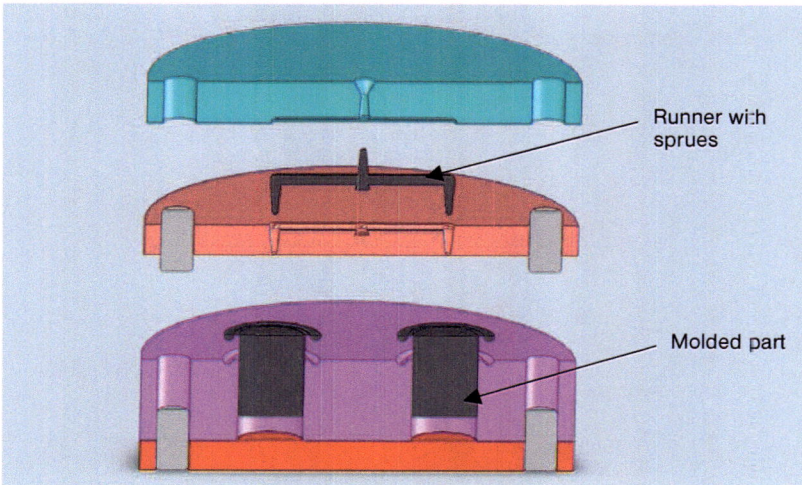

Figure 10.1 Injection mold

10.1   Injection Unit

The injection unit of an injection molding machine can be thought of as a mechanism to feed raw/uncured TSE into a mold at a precise dosage and a controlled pressure and speed. In basic terms, the injection unit consists of a plasticizing screw driven by a motor, an injection chamber, and an injection ram. Similar to plastic injection molding, the space, or gap, bordered by the screw flights, the minor diameter of the screw, and the inner diameter of the barrel becomes smaller in the direction of the material delivery, and therefore, compresses the material as it is transported along the screw. The recommended ratio of compression varies depending on the TSE used. A compression ratio of 1.2 : 1 is typical for most TSEs. Low viscosity materials, or LSRs, usually have no compression ratio (1 : 1). Thermoplastics are able to use a higher compression ratio because there is no threat of scorch.

The length-to-diameter ratio is also important. Most TSE screws range between 12 : 1 to 15 : 1 L/D, a lower range than for thermoplastic screws. The rotating screw, or plasticizing screw, feeds uncured TSE into an injection chamber. Backpressure is placed on the screw to "pack" the material as it supplants into the injection chamber. Subsequent to the filling of the injection chamber, a ram forces the "shot" that was displaced by the screw out of the injection chamber,

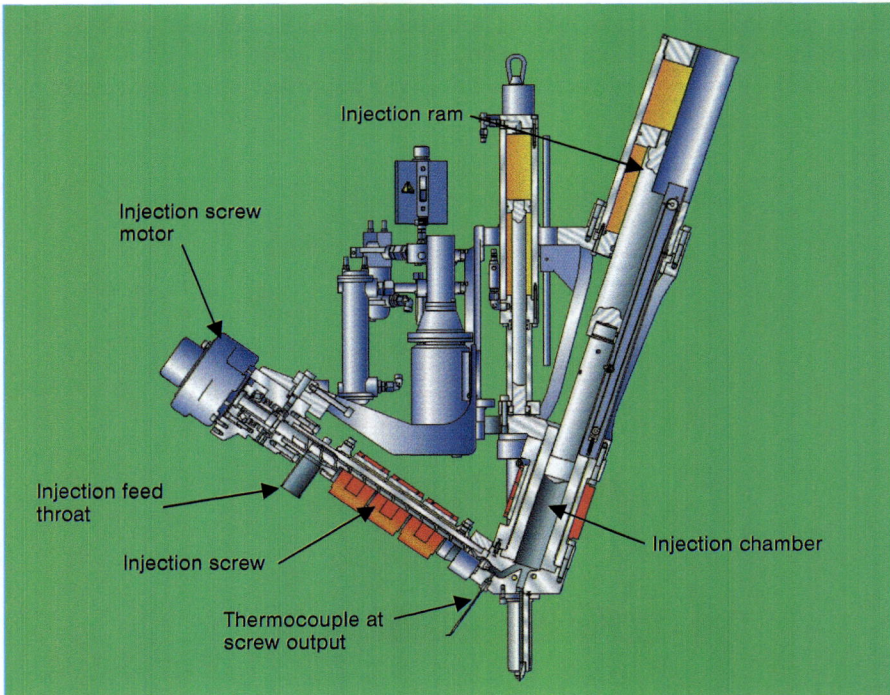

Figure 10.2 "V" pattern injection unit with separate screw and injection chamber (Courtesy of Rep Corporation) [1]

through a nozzle and into a mold. A cushion of material should be left in the injection chamber to apply injection pressure once the cavity is filled. The ram should never be allowed to bottom out during a normal injection cycle.

The injection unit can create tremendous pressure and velocity. It is not uncommon to have injection pressures exceeding 30 000 psi. These pressures are roughly ten times that of compression or transfer molding. An injection flow path with minimal bends and corners is preferred to avoid pressure loss and excessive scorch. Most injection units are designed with the high pressure ram as close to in-line as possible with the nozzle. Since the screw does not feed at high pressure or velocity, it can be designed perpendicular, or at a slight angle to the injection chamber without affecting flow. Figure 10.2 shows a "V"-pattern injection unit favoring the injection ram flow path.

Advantages to a "V" injection unit include:

• Precise injection control by separating the screw from the ram
• The ability to monitor the material's extruded temperature (see thermocouple location in Fig. 10.2)

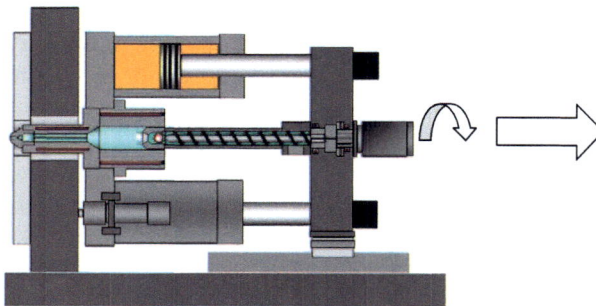

Figure 10.3a Injection unit dosing (filling)

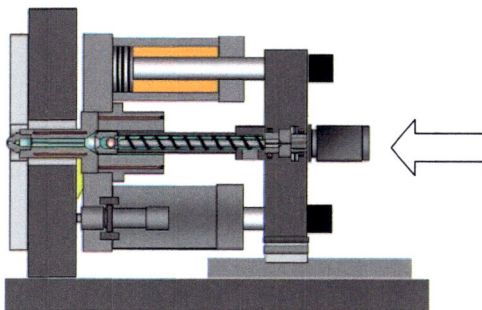

Figure 10.3b Injecting (Courtesy of DESMA) [2]

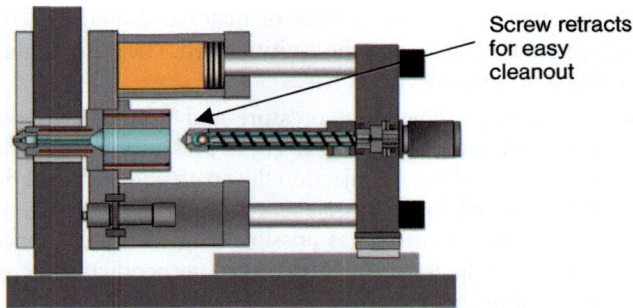

Figure 10.4 Injection screw retracted (Courtesy of DESMA) [2]

A drawback to injection molding is that the uncured TSE is required to travel a long distance and through a multitude of passages before entering the mold. Extensive residence time at elevated temperature can have a harmful effects on physical properties of TSEs. If left stagnant in the injection chamber, TSEs can cure prematurely. This effect varies among TSEs, but avoiding dead spots should be the goal in the design of any injection unit. Improper set-up and operation of an injection unit can lead to cured chunks that can be carried along into the molded article and cause defects. Dead spots in an injection system add complication to material and/or color changes. Extensive purging to change materials can be costly and time consuming compared to compression and transfer molding.

Perhaps not surprising, there are numerous versions of injection units available. Several TSE injection molding machines offer simple clean-out advantages, where the plasticizing screws are simple to remove so that residual TSE can be manually scraped from the screw and barrel. In other cases, the screw can be retracted hydraulically (as shown in Fig. 10.4), exposing the screw for cleaning or inspection. These options may come in handy if the molding machine is intended for frequent material/color changes.

10.1.1   First In – First Out (FIFO)

First in-first out (FIFO) injection units refers to a principle that the material first entering the injection chamber is the first material to exit the injection chamber. In the case of most FIFO units, the screw is integrated into the injection ram and reciprocates. The screw feeds material into the injection chamber — an area directly in front (and in-line) of the screw vacated by the reciprocation of the screw. The screw in this case is also used as the ram to inject the material (see Fig. 10.5). This offers several advantages:

- Avoids "dead" spots, or stagnant material
- Avoids turns, which could create pressure drops
- Utilization of less space
- Easier material and/or color change

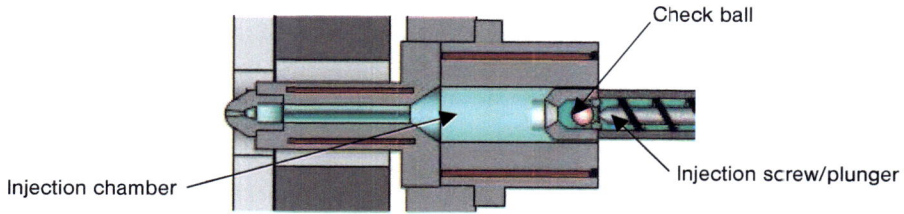

Figure 10.5 FIFO injection unit (Courtesy of DESMA) [2]

10.1.2   Plunger Unit

Plunger-type injection units (Fig. 10.6) were among the first injection units available and are still in use today. In this case, there is no screw. The plunger retracts and allows material to enter the barrel, and then pushes the material into the mold. The advantage of a plunger-type injection system is its simplicity with less moving parts (less wear). In addition, a plunger does not add nearly as much shear stress as a screw. This may be critical for certain materials and/or molded articles.

10.1.3   Injection Controls

The injection chamber, screw, and nozzle typically have temperature controlled zones to control shot-to-shot repeatability. These temperatures can be varied to offer the best flow without threat of scorch. The injection cycle can be controlled by speed or pressure — one is the result of the other. If a specific injection pressure is desired, speed will become the variable; if a specific speed is desired, pressure will fluctuate to achieve that desired speed. The opposing parameter can be used as an alarm, or compiled for SPC data. For instance, if a desired injection speed is chosen, a given pressure should achieve that speed. However, if

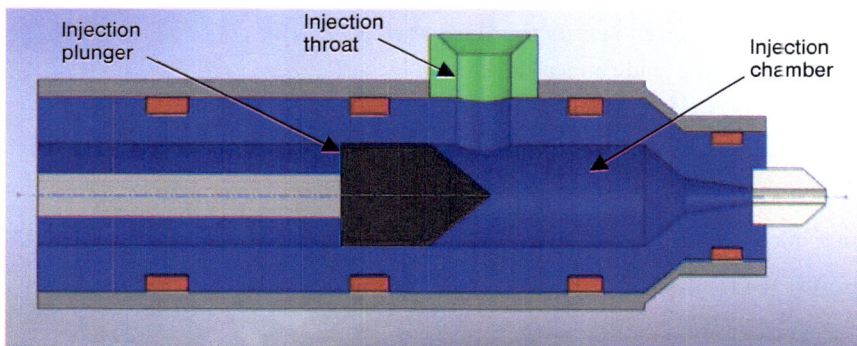

Figure 10.6 Injection plunger unit

the material's viscosity is significantly higher, more pressure will be required to attain the set speed. This spike in pressure could activate an alarm requiring the operator to investigate.

The injection shot size is predicated upon volume. As the TSE displaces the injection ram, a predetermined volume signals the screw to stop feeding material into the injection chamber. The injection unit is then ready to inject.

The injection cycle can be programmed to have several injection stages with different speeds or pressures. These injection stages allow the cavity to fill as quickly as possible without threat of scorching. They also allow for adequate venting and degassing, and packing of the cavity.

Timing of the screw feed is fairly important. During long cure times, the screw should not be immediately activated to fill the injection chamber. This action will only lead to higher residence time at elevated temperatures. The screw feed should be delayed to compensate for the amount of time it takes to transfer an entire shot size into the injection chamber. For example, if it takes ten seconds to fill the injection chamber, the screw feed should commence ten to fifteen seconds before the injection cycle is initiated. It is important to delay refilling the injection unit until late in the cure cycle to avoid lengthy residence time.

10.1.4   Injection Location

Most plastic injection molding machines are designed in a horizontal platform with an injection unit centered on the stationary half of the press. This same basic design is available for TSE injection molding as well. However, many TSE injection molding machines are still vertical platform machines. Usually, vertical machines are equipped with an injection unit centered on the top bolster. In rare cases, some machines are available with injection units that are fixed to the bottom bolster or with side-inject (parting line injection).

Reasons for alternative injection locations vary. Below is a list of the most common rationale:

- Top inject with vertical platform press

   - Insert molding
   - Manual demolding
- Side inject with vertical platform press

   - Minimal distance to cavity (minimal waste)
   - Dual shuttling cold runner blocks
   - Two-shot molding
- Bottom inject with vertical platform press

   - Shuttling or sliding top plate
- Center inject on stationary side with horizontal platform press

   - All-rubber molding
   - Minimal labor to demold, or automatic molding
- Top inject on stationary side with horizontal platform press

   - Two-shot, or co-molding requiring additional injection unit

10.1.5   Material Feed – Stripped

Uncured TSE is usually fed by preformed strips into the feed throat of the plasticizing screw with material that was formed by an extruder or stripped from a mill. The strip is inserted into the feed throat and held against the screw by an operator. As the screw turns, the flights of the screw pull the strip into the injection chamber. Once the strip is pulled _nto the screw, the strip is self-feeding. The TSE strip is often coated with fine powder or a low melt composition, commonly called slab-dip that melts at the milled/extruded temperatures, but solidifies at room temperature. This coating keeps the TSE from sticking together during material handling. However, care should be taken to make sure the molded product is compatible with the coatings and that the coatings do not hinder moldability. Ground TSE pellets can also be used if the uncured TSE does not stick to itself, although pelletized feed is rare.

10.1.6   Material Feed – Stuffer

Many materials, such as gum silicones, do not have the green strength or consistency to feed the uncured material into the screw without breaking. In this case, a stuffer is used where a cylinder is mounted on the screw barrel with a hole aligned with the feed throat of the screw. Material is usually rolled on a mill into a cylindrical shape or "pig" and placed inside the cylinder. A ram forces the material into the throat of the screw, eliminating the potential for strip break.

10.2   Materials

Most TSE materials can be injection molded. However, lower viscosity materials offer distinct advantages due to the lengthy path required in the injection process. Generally speaking, Mooney viscosity exceeding 80 should not be considered for injection molding. Lower viscosity TSEs offer an incrementally greater processability for injection molding. Also, materials should have adequate scorch safety. Since residence time may be substantial during injection molding, retarders in conjunction with accelerators may be required to keep the material from curing prematurely and activate quickly once in the mold.

10.3   Automation

Consistency is the key to success in molding TSE articles. A completely automated system has no operator influence, so cycles are consistent. As any mold opens to remove articles, mold temperatures immediately start to drop. Inconsistent mold temperatures are one of the largest contributors to scrap. Even in automated processes, the mold still needs to open and temperature will fluctuate. However, consistent fluctuation is controllable.

Automating TSE molding presents plenty of challenges. The handling of uncured TSE often warrants the need for an operator in compression and transfer molding. Injection molding, at least, minimizes concerns regarding the handling of uncured material, since material is automatically fed through the injection unit and ultimately into the mold. Automation is

MOLD TEMPERATURES DURING MOLDING CYCLES

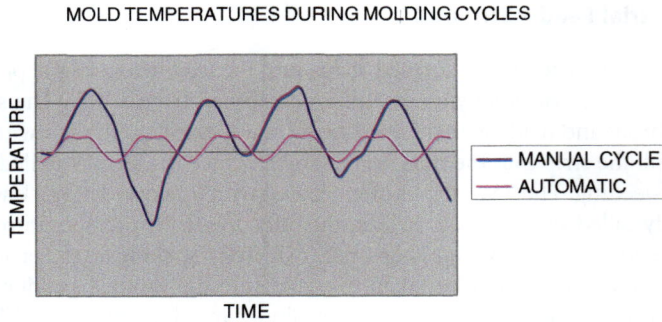

Figure 10.7 Temperature fluctuations

best suited for all-rubber (non-inserted) articles molded in a horizontal injection molding machine with an insulated delivery system (such as a cold runner). Part geometry of the molded article also plays a significant role in its ability to be automated. Articles with deep draws or undercuts invariably need an operator to remove them from the mold. Articles that are easily removed by air or brushes are good candidates for automation.

Conventional injection molding of multi-cavity molds requires a runner system that cures along with the molded article. This runner system most often requires an additional parting line — often referred to as a three-plate mold. The runner is attached to the main sprue as well as the secondary sprues that feed into the mold cavity (see Fig. 10.1). Plastic injection molding machines use robots (pickers) to automatically remove this runner system. However, plastic is rigid and is easily removed as a single unit. TSE is elastic and tears easily, especially when hot, and therefore is difficult to remove intact. A cold runner, or better yet a valve gated delivery system, is preferred if automation is the goal. These delivery systems could negate the need for an additional runner parting line, and may eliminate the need for the removal of a gate or sprue entirely. If so, demolding can be isolated to the molded article. The less material needed to be removed from an injection molding process, the less complicated automation becomes.

Flash removal is a big concern in automation. For injection molded processes slated for automation, consider hardened individual cavity inserts similar to those described in flash-less transfer molding. These cavities should be spring loaded with heavy disk springs to keep the parting lines closed during injection. Flashless grinds can be added to vent cavities and also avoid flash. Since injection pressures are much higher than with other molding methods, higher spring pressure may be needed to avoid flash from injection. Eighty percent of total clamp pressure should be distributed equally among all cavities as a designated spring force. Careful consideration to land widths is recommended. Heavy spring force can damage narrow lands and grinds. Typical flashless transfer inserts which are subjected to no more than 3000 psi may get destroyed under typical 20 000 psi cavity pressure. Cut-in-plate tools are not recommended for automation, since flash may be difficult to control.

Once the multitude of conditions hindering automation have been removed, or reduced, a method of article demolding needs to be determined. Robots with grippers are an effective method of removing articles. Custom gripper ends may need to be fabricated to fit exposed geometry. Sensors can be incorporated into the gripper that determine if an article is actually held in the gripper once demolding movements are completed. If a part is not detected, a signal can shut down the press movements, alarming the operator that an article may be left in the mold.

Pressurized air strategically positioned to blow parts from the mold is an effective method of part removal. Very high air pressure can be used to assure article and flash removal. OSHA limits the amount of air pressure in situations where an operator can come in contact with the air stream, so guarding may need to be established to avoid operator contact.

Rotary brushes are an aggressive yet effective method of removing articles and flash from molds. This is best performed on a horizontal injection press. A motor driving a rotary brush is attached to the end of a robot. After a mold is opened, the robot is fed down along an entire parting line of a mold while the brush is turning, forcing the articles and flash from the mold (see Fig. 10.8). Pressurized air can be added to the robot and aimed at the parting line, while the brush is rotating to supply additional insurance that the articles and flash are adequately removed. An automated mold release sprayer can also be added to the robot, releasing a controlled amount of mold release onto the mold parting line as the robot retracts. For a more aggressive approach, paddles instead of brushes can provide greater force in removing articles from the mold. Whether brushes or paddles are used, materials should be non-abrasive to avoid damage to articles, mold coatings and mold cavity surfaces. Some experimentation may be required to assure repeated article removal without damage. Brushes or paddles are definitely wear items and need to be added to a strict preventative maintenance program.

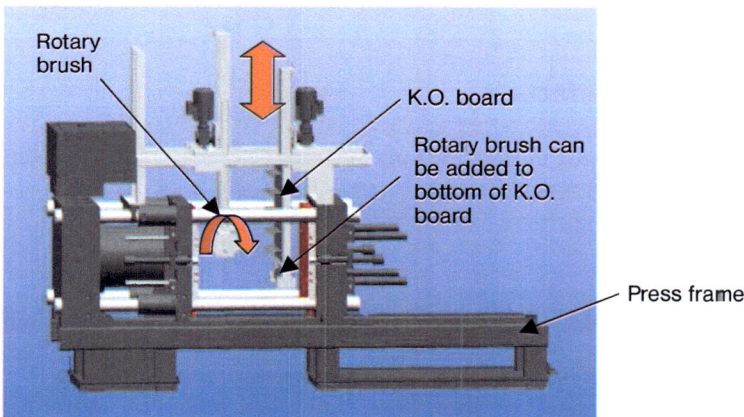

Figure 10.8 Automated brushing system (Courtesy of DESMA) [2]

10.4   Mold Construction

Since injection molding can produce extremely high injection pressures, mold plates need to be sturdy enough to withstand these pressures and avoid bending and flexing. This is of principle importance regarding the top mold plate. Open-mold purging can be particularly dangerous if the top clamp plate is not of adequate substance. The top clamp plate should be pre-hard steel, 3″ thick at a minimum. Plates, other than the top clamp plate, should be a minimum of ¾″ thick.

A variety of mold designs and constructions will work for injection molding. Simple cut-in-plate molds work fine in many applications, if an operator is used. Since cut-in-plate molds tend to flash, the flash is removed simultaneously with the molded article.

Flash can be more difficult to control in injection molding versus transfer molding, with compression molding typically being the most difficult to control. Force is the product of pressure and surface area. Since injection molding can have cavity pressure up to ten times higher than transfer molding, given the same cavity area, injection mold parting lines may overcome clamp force, while transfer mold parting lines may not. Spring loaded inserts as described earlier can be overcome by excessive injection molding pressures. It is best to control flash in injection molding by controlling the shot size and injection pressure.

Minimum mold height and maximum daylight need to be considered when designing a mold. An adequate amount of daylight is required for article removal, mold cleaning, and insert loading, if required.

An advantage of injection molding is that raw material feed is performed by the injection molding machine. This limits the need for an operator. For parts that are easy to remove from the mold while flash is kept to a minimum, automation can be administered by use of robots and brushes.

10.5   Molding Defects

Injection molding can cause most of the same types of defects as transfer molding. However, some defects are more prevalent in injection molding.

10.5.1   Scorch

Since the TSE is required to travel a much longer distance, and at an elevated temperatures, scorch is common in injection molding. Compounds can be altered to supply greater scorch safety. Scorch can be minimized through temperature reduction in the mold and/or injection unit, a reduction in injection speed, a reduction in screw RPM, and/or a reduction in backpressure.

10.5.2   Cured Stock

Cured stock can come from sprues or runners that were not entirely removed from the prior heat. The plasticizing screw or injection chamber could have a dead spot, or runaway heater, or an excessive screw RPM where cured TSE can get carried along into the molded article.

10.5.3   Adhesive Wash

If bonded inserts are used, sprues or gates should be located away from the insert surface to be bonded. All the material that needs to fill a cavity has to pass through sprues or gates. At high velocity, the material may wash away the adhesive from the inserts. This can result in poor bond and mold fouling.

10.6   Injection Transfer

Injection transfer simply introduces an injection unit to a transfer mold. The same principles of transfer molding apply, except an operator is not required to load TSE preps on top of the mold. In this case, the TSE is injected into the pot prior to the press closing. Advantages include preheated TSE with better flow and potentially quicker cure, TSE usage flexibility, less operator influence, and faster cycle times with the elimination of prep loading. Injection transfer may also lend itself better to automation since an operator is not required to place preps on the mold.

Wasteless transfer molding, as described in the transfer molding chapter, is an excellent candidate to convert to injection molding. These processes can also be run in both vertical and horizontal machines. Automation can be included by capturing the sprues in a steel plate that utilizes a knock out board or brushing system to remove the sprues from the plate. Fabric material can also be advanced automatically through an injection machine. In this case, a roll of fabric is fed through the machine between the insulator board and the top insert. As the sprues get captured and advance, the fabric is discarded along with the sprues

10.7   Injection Compression

Large parts are best suited for injection compression molding. With this method, TSE is injected into an open mold, and the press simply closes on the material similar to compression molding. In essence, injection compression molding is compression molding that places the material into the open cavity via an injection unit.

In cases where sprue vestiges are not allowed on the part, the sprue can be located near an edge of the cavity and angled toward the actual cavity, or in the center of a donut-shaped part.

10.8   Cold Runner Injection

For injection molding, insulating the runner system to a temperature well below the cure kick-off has become very popular because the material

- can remain in the insulated state for a period of time,
- provide a delivery path to the end product, and
- not be discarded during the molding cycle.

Typically, a manifold with a predetermined runner pattern is utilized with adequate cooling lines on the injection side of the mold (see Fig. 10.9). The mold (containing the cavity) is still heated to its normal temperature. Between the mold and the cold runner manifold is an insulator board. Traditional cold runner manifolds incorporate a cooled nozzle that protrudes through the insulator board into the heated portion of the mold. The molded article still requires a sprue, and once the part is cured and the mold is opened, the part is ejected and the sprue is removed. The sprue, in this case, breaks off inside the nozzle where the cooled portion prevents cure kick-off. Cold runner injection molding can dramatically reduce the amount of material discarded.

The cold runner manifold is cooled by liquid. This can be treated water, glycol, or oil. The liquid is routed through the manifold and into the tips of the cold runner nozzles. The liquid passages through the nozzle can easily become clogged. Therefore, the nozzles should be routinely disassembled and cleaned.

Similar to conventional injection runner systems, sprues should be equally spaced apart so that flow is identical among all cavities. Likewise, cooling jackets and lines should provide equal cooling to each runner leg and nozzle. Temperature can affect the viscosity of the TSE. If a hot spot exists in the cold runner, the nearest nozzle may allow its cavity to fill faster than adjacent cavities. Also, hot spots can cause TSE to cure inside of the cold runner, blocking material flow.

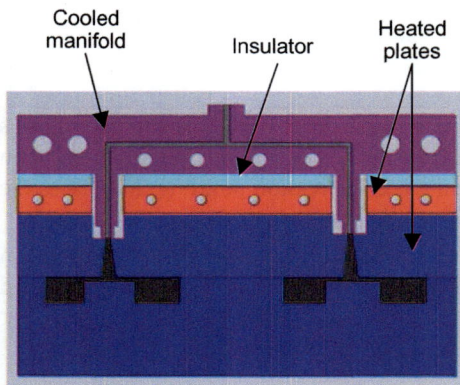

Figure 10.9a Injection cold runner (closed)

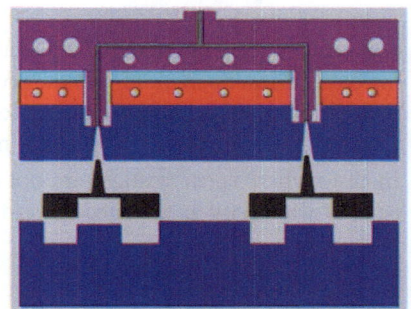

Figure 10.9b Injection cold runner (open)

As with any runner system, sharp corners should be avoided. Runners within a cold runner are usually gun-drilled through the block. Plugs are inserted to divert material flow. These plugs should be precision-machined to include radii at flow intersections to avoid sharp corners and dead spots in the flow channel. The plugs should be keyed for anti-rotation.

Some tool makers prefer split cold runner blocks so that corners can be easily blended, negating the need for gun-drilling or plugs. The split, or parting, line occurs at the center of the runner. Each half of the cold runner is milled with a half-round runner and when assembled completes a full round runner. Split cold runners require a tremendous amount of large high-yield bolts spaced in close proximity to each other and the runner, to prevent the injection pressure from prying the parting line open.

Invariably cold runners will cure up and TSE will crosslink in the runner channels. This can be a real challenge to remedy. Split cold runners can be unbolted and cured stock can be removed. For solid cold runners, plugs need to be removed and the cured stock needs to be drilled out. In either case, it can take several hours to clear a plugged cold runner, and it may even be necessary to send it to outside services to be cleared.

Cold runners require special procedures to shut down. Crash cooling is a term used when the mold is to be shut down and the cold runner goes into a cycle of flushing constant cold coolant through the manifold until adequate cooling takes place in the mold so that heat will not soak into the cold runner block. This crash cool prevents the TSE from curing inside the cold runner. Another technique to prevent curing upon shutdown is to purge identical polymer without any curative through the cold runner. In this case it does not matter if too much heat reaches the cold runner block, because the TSE inside the cold runner will not cure. Upon start-up, the material with curative is purged through the cold runner block clearing any residual non-cure compound. Since the base polymer is identical, contamination is held to a minimum.

10.9   Valve-Gated Cold Runner

The plastics industry has utilized valve gates for many years. Recently, the TSE industry has advanced to the same technology. Valve-gating for TSE molds involves a cold runner similar to the one used in cold runner injection molding. Instead of using a cooled nozzle with a hole in the center, a nozzle is made with a pneumatically operated pin with a tight fit shut-off at the cavity end. Figure 10.10 shows a cross-section view of a typical valve-gated cold runner injection mold. The valves can be activated independently (in multi-cavity molds). In valve-gated molds, zero waste can be achieved, and fill control can be easily adjusted, rendering valve-gated cold runner molding among the most economical molding methods available.

A valve-gate nozzle is similar to the nozzle in a conventional cold runner. It requires water jackets to keep the TSE inside of the nozzle from curing, and intersects a runner which feeds uncured TSE through the center of the nozzle. What is different is that the nozzle has a pin in the center, which is activated by a pneumatic signal that retracts the pin, allowing a path for the uncured TSE to fill a cavity. This pin can be activated independent of the injection cycle. This means that if, with all good intentions and design considerations, flow is

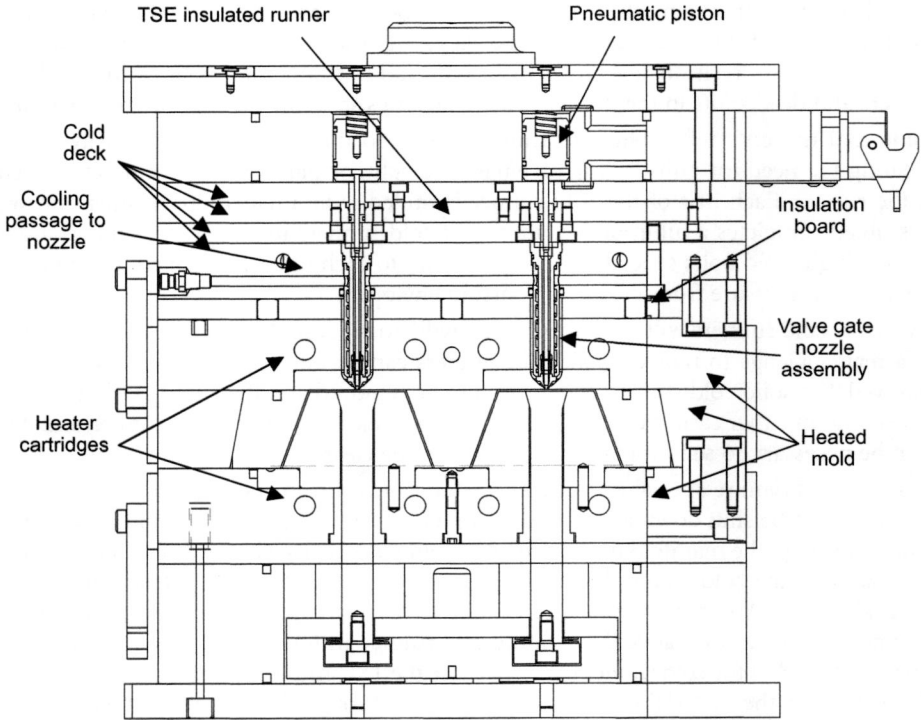

Figure 10.10 Cold runner valve gate mold cross section (Courtesy of MR Mold) [3]

not consistent from one nozzle to the next, the pneumatic valve can compensate by leaving certain valve gates open longer than others to allow for equal fill. Valve gates can be considered flow compensators.

Valve-gating is capital intensive and works best with very forgiving materials with low viscosity. Gum silicone and LSRs are ideally suited for valve gate technology. Valve-gating does require ample space. Valve gate nozzles are fairly large in diameter and limit cavity spacing.

10.10   Injection Pressure Considerations

Injection pressure can easily overcome clamp pressure. Considering ultimate injection pressure at 30 000 psi, assume a pressure loss through the injection unit, nozzle, runner, sprues, etc. to reach 50 %. This pressure should be multiplied by the total molded surface in the mold (all surface areas including potential flash and overflow grooves should be considered in the surface area calculation), and should be less than 75 % of total clamp force. Even at the pressure loss indicated above, it does not take much molding surface area to overcome a typical clamp force. Moreover, assuming that all platens, bolsters, insulator boards, mold

Figure 10.11 Valve gate nozzle (Courtesy of MR Mold) [3]

plates, and cavity inserts are perfectly parallel is unrealistic. Press parallelism and insulator board exchange are of considerable importance in injection molding.

10.10.1 Pressure Compensator

In some cases it is near impossible to fill multi-cavity molds in perfect unison. Even in assumed perfectly balanced runner systems, one cavity may start to flash because the cavity is already filled, while an adjacent cavity is still filling. In these cases, the injection shot size is predicated on the least filled cavity. In other words, quicker filling cavities may flash to compensate for the slowest and least filled cavities. At 10 000–20 000 internal psi, it is difficult to keep cavities shut under any design consideration. High pressure spring loaded pins can be used as a buffer device. The theory is that these pins are designed to collapse under injection pressure. Once the injection cycle is complete, the cavities are not completely filled and the compensating pins release their sprung energy and continue to fill the cavities at a much lower pressure. This pressure should be designed at 2 000–3 000 psi, more in line with compression or transfer molding pressures. The pins need to be in very close proximity to the cavity, or directly in the cavity, to work effectively. PTFE seals should be used with the pins to prevent TSE from bleeding down the pin. In some seal applications, this pin can be incorporated into a trim cap (or sometimes referred to a maiden-head).

This approach can also be used in valve-gate technology. Typical valve-gated nozzles operate the valve gate pin pneumatically to open and close the gate. Disk springs at a designated pressure can be used to force the gate shut. In this case, pneumatics is only used to overcome the spring pressure to open the valve, and the spring force closes the valve and simultaneously acts as a pressure compensating pin as described above.

10.11   Conclusion

Injection molding is a complicated process with many potential variations. Injection molding should be utilized for high-volume applications, or several moderate-volume applications with identical materials.

Injection molding offers the best process control set-up of any molding technique. Its material handling sets the stage for automation. When combined with an insulated delivery system, injection molding can yield very cost-competitive processes. Injection molding capability is a must for every custom molder.

References

1. Rep Corporation, Bartlet, IL.
2. DESMA, USA, Hebron, KY.
3. MR Mold, Brea, CA.

11   Liquid Silicone Rubber

LSR (liquid silicone rubber) is a very low viscosity silicone; pourable (or self-leveling) in consistency. This same material is also referred to as LIM (liquid injection molding), and is a registered trademark of Momentive Performance Materials Holdings, Inc. This book will use the term LSR when referring to this low viscosity silicone. LSRs range in viscosity from 300 000 to 1 000 000 cps, as compared to typical gum TSEs (including millable silicones) with a viscosity of 5 000 000 to over 20 000 000 cps [1].

When LSRs were originally introduced, they had difficulty competing against their millable silicone counterparts. Materials were more expensive, physical properties were inferior, and processing offered only marginal gains. Recent improvements in physical properties and enhanced processability escalated LSR into a competitive position with millable silicone. Faster cure times and lower viscosity (which tends to offer less scrap and flash free products) not only staged LSR competitively with millable silicone, but other TSEs, such as natural rubber.

This chapter will describe the LSR fluid delivery system, press injection units, tooling, materials, and two-shot molding.

Figure 11.1 LSR pumping unit (Courtesy of Fluid Automation, Inc.) [2]

11.1   The System

LSR is purchased in drum form, usually in 5 or 55 gallon containers. The LSR molding process involves a two-part system consisting of part A containing a platinum catalyst and part B containing a silicone hybrid crosslinker and an inhibiter to afford reasonable mixed pot life [1]. Other than crosslinking additives, both containers house the same silicone.

Part A and B drums are placed near an injection molding machine, see Fig. 11.2. Pumps, similar to grease pumps, deliver the material at room temperature and at a 1:1 ratio (at approx. 80 psi) to the static mixer. Some adjustment (up to 5 %) can be made to the ratio without adversely affecting the process. Slightly adjusting this ratio is typically done to empty both barrels simultaneously in the event that one barrel will empty prior to the other. LSR is purchased as a kit (part A and part B). If one barrel has remaining material, it will be discarded, so slight adjustments may minimize left over material. Variable ratio machines are available offering the molder the ability to fine-tune hardness where part A and B have different hardness.

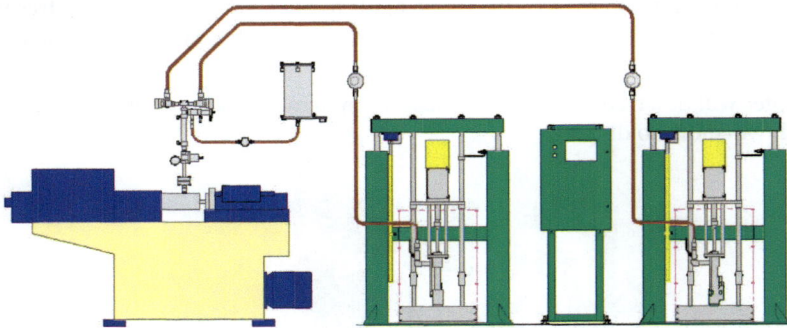

Figure 11.2 Typical LSR processing schematic (Courtesy of Fluid Automation, Inc) [2]

11.2   The Static Mixer

The static mixer combines part A with part B by dividing and shearing the stream of material repeatedly through a series of mix elements (see Fig. 11.3). Static mixers vary in size and the amount of elements (turbulent flutes). A 32-element mixer is typically adequate for most applications where a plunger injection unit is used in conjunction with a color pigment, whereas a 20-element mixer will suffice when a screw injection unit is used. Mixer sizes range from 3/16″ to 1″ in diameter and vary in length, depending on the shot size requirements [3]. In some high shot size applications, where a screw injector is used without color pigment, the static mixer may be eliminated.

Once combined at the static mixer, the material becomes active. Residence time in the mixed condition should to be held to the bare minimum. Therefore, sizing the static mixer to minimize residence time is important. As with any TSE, cooling the active mix temporarily retards the crosslinking process, and therefore the path between the static mixer (including the static

Figure 11.3a Static mixer
(Courtesy of Fluid Automation, Inc.) [2]

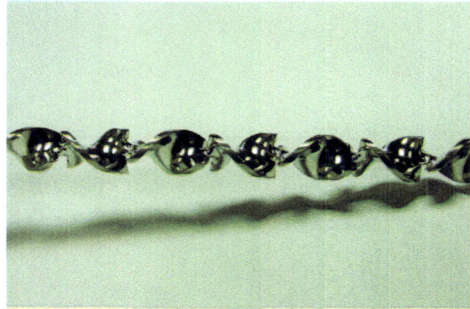

Figure 11.3b Static mixer enlarged view

mixer) and the press injection unit is temperature controlled. The distance from the static mixer to the press injection unit should be minimized as much as possible. More importantly, the volume of material should be minimized.

One pumping unit can feed several presses simultaneously. Pressure loss and feed rate variation at each individual press can result from assigning too many presses to a single pump. This pressure loss variation is sometimes overcome by adding a secondary (feed) pumping system incorporated in-line just prior to the static mixer. Utilizing a secondary pump is referred to as a transfer and meter system. Advantages to transfer and meter include:

- Allowance for several presses to use the same main pumping unit without threat of pressure loss if several presses draw material simultaneously,
- A more accurate pressure and dosage to feed the injection unit,
- An additional point of purge, and
- Lower up-front cost by sharing pumping units.

Figure 11.4 shows a schematic of one pumping unit feeding two presses; one with a color pot, one without.

For high volume applications, where downtime is critical, two pumping units can be incorporated. In this case, one barrel can be changed while the other continues to offer material to the presses. Figure 11.5 shows an automatic changeover feed system.

11.3   Injection Unit

The exit path of the static mixer leads to the injection unit of an injection molding machine. A plunger injection unit is usually not recommended for most TSE molding. However, since the LSR can be adequately mixed in the static mixer, a screw injection unit may overwork the LSR. In fact, many LSR injection molding machines are equipped with plunger injection units. A plunger unit is simple, accurate, and easy to maintain and clean. Sealing a plunger unit is much easier than a screw, because the plunger only sees reciprocating motion as opposed

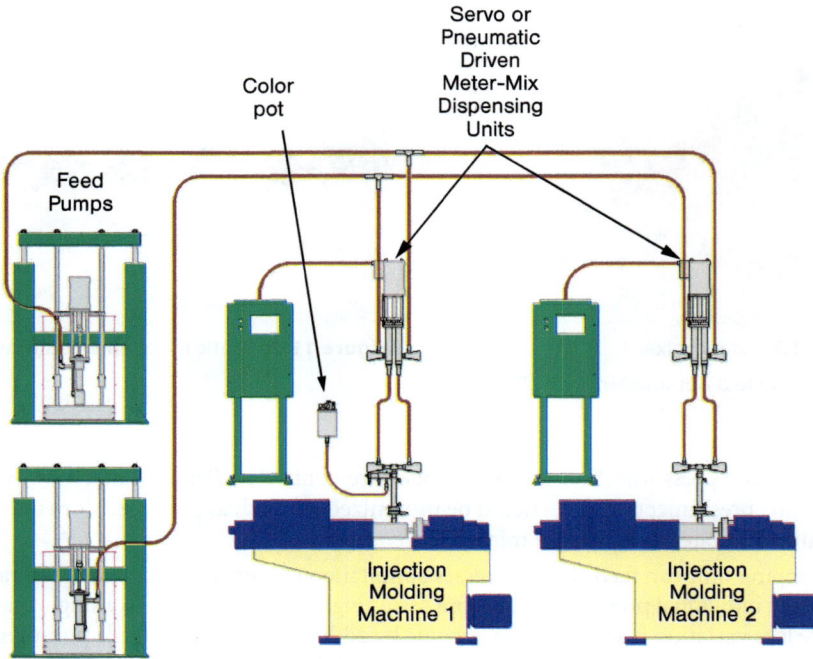

Figure 11.4 Transfer and meter system (Courtesy of Fluid Automation, Inc.) [2]

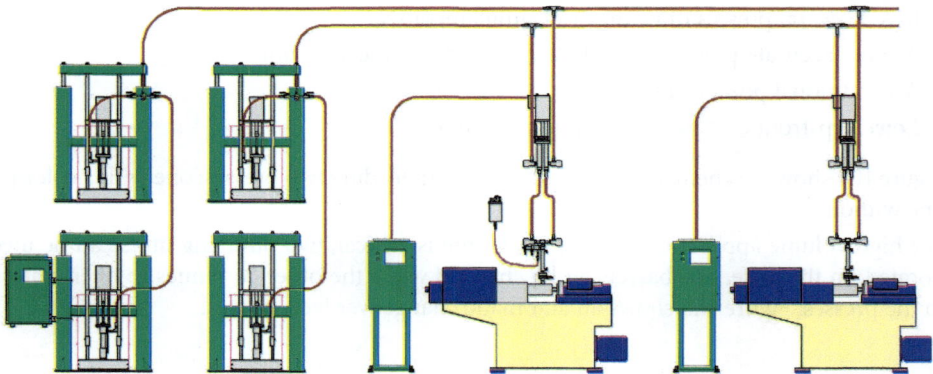

Figure 11.5 Transfer and meter with automatic change-over feed system (Courtesy of Fluid Automation, Inc.) [2]

to rotational and reciprocating motion from a screw unit. The plunger will last much longer and seal much better than a screw. Also, the pumping unit supplies adequate pressure to fill the injection unit, so a screw is not necessary. The injection unit must also be temperature controlled.

Low viscosity allows injection pressures for LSR to be much lower than their millable counterparts. While millable TSEs may require 20 000 psi injection pressures to adequately fill cavities, LSR injection pressures may only require 100 to 5 000 psi, although higher pressure is available if needed.

11.4   Molds

LSR molds have some similarities to other TSE injection molds. LSRs are exceptional candidates for injection molding because they have such a low viscosity. Required injection pressures can be much lower than millable TSEs, and therefore tend to provide less potential flash on molded articles. Additionally, clamp pressures can be much lower, so press size can be much smaller for articles molded from LSR.

LSRs are prone to flash, but fortunately lower injection pressures help minimize this condition. Mold construction should take into account that parting lines may flash at gaps as small as 0.000 2″. Therefore, high quality, precision mold construction should be practiced when building an LSR mold. LSRs exhibit poor heat tear resistance, so deep undercuts and convolutions requiring the article to stretch during demolding are cause for concern. Parting lines often have sharp corners that may propagate a tear. For LSR molds, it is better to slightly break the sharp corner of parting lines and allow for a witness line as opposed to a minimal witness line with frequent hot tears.

Unique among TSEs are LSRs potential requirements for transparency and cosmetic appearance. Mold surface finish plays a big role in the amount of transparency exhibited in the molded article. Mirror polished cavities (SPI, A-1, or A-2) produce the highest degree of transparency, but also create an affinity for the LSR, which translates to greater demolding forces. This condition is not uncommon among TSEs, although mirror finishes are usually not required for cosmetic appearances for non-LSR articles. An SPI D-2 or D-3 bead blasted finish offers the best demolding results.

Valve-gated cold runner systems are very popular with LSR injection molding. Low viscosity materials, such as LSR, make valve-gating an economical method of molding Valve-gating in conjunction with flash-free molding allow LSR processes to run fully automatic without sprue waste, and can therefore offer very competitive pricing for molded articles.

LSRs are not recommended for compression molding. Compression molding requires a high-viscosity material to "pack" the cavity. However, transfer molding LSR can be done by purging material from the pumping unit (just after the static mixer) onto a special carrier and offering the material to the transfer pot. A paper sheet can be placed onto a scale under the purge discharge, an accurate weight can be purged onto the paper, and the combination can be placed in the pot with the LSR material side facing the mold.

A better approach is to consider injection transfer molding LSR. Injection transfer avoids the difficulties related to the material handling for transfer molding LSR. This process requires the same pumping unit and press as used for injection molding, except that a pot is used instead of a runner system. To effectively transfer mold, the material is injected into a slightly opened pot — closed enough to engage the piston into the pot to prevent leakage, and open enough to freely fill the pot. Once the pot is filled, the press is closed to pressurize the pot and force the LSR through sprues into the cavity.

Utilizing a cold pot with injection transfer molding of LSR can be accomplished as a wasteless system as described in the transfer molding chapter. Wasteless transfer molding can offer flashless articles, high cavitation, and minimal material waste in a fairly forgiving process. For high-volume applications, this process may offer the most competitive option.

11.5   Materials

LSR is available in a hardness range of 15–80 (shore A). Once parts A and B are combined and heat is applied from the mold, crosslinking is quickly initiated. LSRs predominantly use a platinum cure system, which offers cure times that are counted in seconds as opposed to minutes with most millable TSEs. LSR is usually crosslinked at 375 °F, resulting in t90-times that are two to three times faster than peroxide-cured millable silicone at 340–355 °F [4].

Millable silicones can use platinum cure systems, but the entire cure package may vary slightly from the one used in LSRs. Discounting the cure system, chemically LSRs can be considered identical to their millable counterparts, with the only difference being molecular weight, resulting in differences in viscosity and some physical properties.

LSRs, as with all TSEs, are considered shear-thinning materials (pseudoplastic). An LSR with a base viscosity of 700 000 cps may exhibit a viscosity of 50 000 to 100 000 cps at the point of injection due to the stresses created from material flowing through gates, runners, cold runners, etc. [4].

It is important to note that platinum cure systems are prone to breakdown by contamination, particularly by sulfur or sulfur-donor cure systems that are predominantly found in most organic TSE compounds. For this reason, platinum cured materials should be segregated from any sulfur borne compounds from storage, to mixing, through molding. HVAC units should not be ducted to allow air to cross from sulfur bearing operations to platinum bearing operations. Molds that come in contact with sulfur-based TSEs should be cleaned thoroughly prior to processing with LSR materials.

LSR molding utilizes many hoses to deliver LSR from the pumping unit to the mold. Rubber hoses that are produced with sulfur curatives should not be used with LSR. It is best to use PTFE lined flexible hose for LSR molding to avoid potential contamination with sulfur. In addition, careful attention to seals and gaskets should be exercised to avoid any potential contamination. For critical medical applications, stainless steel should be used for all plumbing and valving, including the injection unit, where contact with LSR may occur.

Since no compounding is required by the molder, many LSRs are certified to meet industry standard specifications. These standards include ASTM, FDA, Rohs, pharmacopia, ISO, NSF, UL, and others. Using materials that already meet regulations offers faster development and testing of LSR molded articles.

11.6   Special Applications

11.6.1   Medical

Good physical properties, chemical resistance, heat resistance, high transparency, ease of sterilization, neutral odor/taste, no by-product formation, and most importantly biocompatibility (Tripartite Biocompatibility Guidance/US Pharmacopoeia 23 Class 6 Plastic) places LSR in a category favorable to the medical device market [4].

LSR parts can be easily sterilized with minimal effect on physical properties. EtO, gamma/electron beam irradiation, steam autoclaving, or dry heat are common methods used to sterilize LSR parts [4].

11.6.2   Food Contact

Materials can also be compounded according to FDA-CFR 177.2600. Very low extractable volumes, volatility levels, and no nitrosamine content make LSR products attractive for baby bottle nipple production, special feeding devices, and milk contact components [4].

11.7   Color Or Other Additives

LSR has an advantage of adding color at the press. Since the base material is clear, vibrant colors can add aesthetic appeal to the end product. Color pots are tapped into the input side of the static mixer. Sequenced by material delivery signals, a valve is activated releasing a precise percentage of color pigment into the A-B mix. The percentage of color is adjustable to meet specific requirements. The color pigment will mix through the static mixer and incorporate into the A-B mix. Changing colors is simply a matter of changing a color pot (or cleaning it), thoroughly cleaning the static mixer, and purging the injection unit.

Other ingredients can be added using the color pot, such as antimicrobials, or other ingredients that can enhance the material's resistance to specific environments.

11.8   Material Change

Material change in LSR molding can be messy. If pumping units can be dedicated to a given material, the only concern in barrel changing is contamination, or cured material beyond the static mixer. In other words, timing is important for barrel changes. However, if an entirely different LSR is to be used in a pumping unit, the entire system needs to be purged and cleaned. Material changes waste a lot of material, and it is therefore recommended to use dedicated materials for a given system. It is a good idea to have several hoses dedicated to certain materials to prevent extensive purging.

11.9   Similarity to Plastic Injection Molding

Low injection pressures, quick cycle times, low scrap, and predominantly horizontal press platforms places LSR injection molding in a similar arena as plastic injection molding. Plastic injection presses can easily be converted to LSR molding by purchasing a pumping unit and converting a screw and barrel to a thermoset unit. As a result, many plastic injection molders have migrated into LSR molding. Where millable TSEs still fetch good profit margins, many LSR applications have been penetrated by plastic injection molders and, as a result, have become increasingly competitive.

11.10   Two-Shot Molding

Similar to two-shot molding in the plastic injection industry, several press manufacturers have developed two-shot injection presses (sometimes referred to as co-molding) that can deliver thermoplastic in one barrel and LSR in another. Many molding applications require silicone over-molding onto plastic substrates (see Fig. 11.6). Traditionally, these parts required two separate machines: a plastic injection molding machine and a TSE molding machine.

There are several techniques used to perform two-shot molding as far as the mold is concerned. Rotary molds, as used in the plastics industry, use molds that are in essence split into two unique segments on the stationary side (A-side) of the mold — a plastic half and an LSR half; each having its own material delivery system. The moving side (B-side) of the mold turns either the mold or the mold and platen 180° to introduce the just-molded plastic component to the LSR portion of the mold (see Fig. 11.8). In some cases, only a core bar or a plate are rotated (see Fig. 11.7). The finished article is ejected prior to mold closure to evacuate the cavity in preparation to fill the plastic half of the mold.

A second molding method is similar to the rotary method, except the molds do not rotate. Instead, the molded plastic components are removed from the plastic half of the mold by a robot and placed into the LSR half (see Fig. 11.9). Ejection of the finished article can be done by the same robot, or by a traditional K.O. system.

Figure 11.6 Two-shot molded article (Courtesy of Engel) [5]

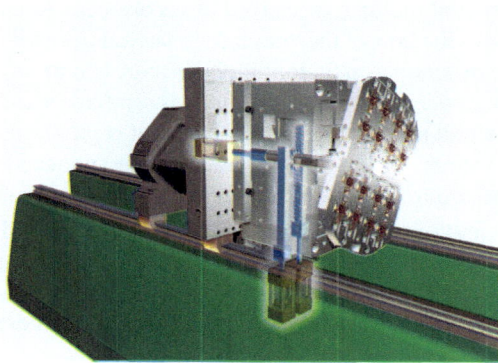

Figure 11.7 Rotating plate in B-side of 2-shot mold (Courtesy of Engel) [5]

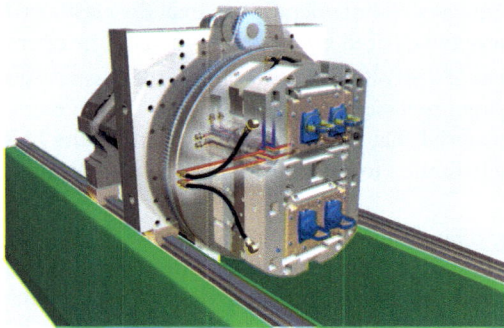

Figure 11.8 Rotating entire B-side of 2-shot mold (Courtesy of Engel) [5]

Figure 11.9 Pick and place robot on B-side of 2-shot mold (Courtesy of Engel) [5]

A third method for two-shot molding is referred to as core-back (see Fig. 11.10). In core-back, a core pin or insert blocks the flow of thermoplastic material from filling into the LSR portion of the cavity. Once the thermoplastic is solidified, the core pin or insert is retracted, and the LSR is injected into the cavity; once the pin or insert is retracted, the LSR meets the solidified thermoplastic. This method is not widely used because it typically extends the cycle time. In either method, a cold runner for the thermoset and a hot runner for the thermoplastic can be included into this set-up to eliminate secondary operations and material waste.

There are still only a few self-bonding LSRs available. Their bondability may be limited to certain substrate types. Where specific LSRs are required that are not available in self-bonding — or are self-bonding, but not to the required plastic — mechanical bonding, if allowed, may be the best option. In this case, holes, grooves, or undercuts incorporated into the plastic design help lock the LSR onto the plastic.

In many cases, mechanical locking features will not suffice in the end product's application. For these cases, an in-process plasma treating process may prove favorable. Whereas most plasma treating to enhance bonding of plastic to TSE is performed in an off-line operation with a low pressure plasma apparatus (vacuum), atmospheric pressure plasma treatment could be incorporated into a two-shot operation to treat the plastic prior to being overmolded. This process may achieve the desired bond without an off-line operation [6].

As with any overmolding of TSE onto a thermoplastic, the plastic requires a high temperature range. Most LSR molding processes operate above 300 °F. Therefore, the plastic substrate is relegated to a plastic material that maintains rigidity during the LSR molding temperature. These plastics are usually limited to engineering grade materials that can be relatively expensive.

Two-shot molding may offer a cure time improvement for the LSR portion of the molding in comparison to two separate operations. In two-shot molding, the plastic is still warm from being molded. Plastic is a good heat insulator, and if dropped into a LSR mold at room temperature, requires time to reach the cure kickoff temperature of the LSR. The LSR will not cure and bond to the plastic until this temperature is reached. With two-shot molding,

Figure 11.10 Core-back design for 2-shot mold (Courtesy of Engel) [5]

the amount of time needed to raise the plastic temperature can be eliminated from the LSR cure time, shortening the overall cycle time. Another advantage of two-shot molding is that the plastic surface is not offered a chance to become contaminated. Just after it is solidified in two-shot molding, it is presented to the LSR portion of the molding operation.

It is strongly recommended to prototype a two-shot operation. In many two-shot applications, the LSR tool-half needs to coin into the plastic substrate to prevent LSR from flashing into unwanted locations. Coining into a warm plastic substrate may have different results versus coining into a room temperature substrate. Bond, thermal expansion, and shrinkage all play a role in successful two-shot molding. Understanding the thermal properties of the plastic and LSR during the prototyping can avoid downstream problems. If bond is critical to the application and self-bonding LSRs are used, extensive experimentations with various temperatures and timing need to be performed. DOEs are critical to the success of marrying the two materials successfully. If a multi-cavity two-shot mold is considered, building only one cavity for each material portion and sampling the mold/process, may resolve design issues before proceeding to production with the balance of the cavities.

Two-shot molding does not need to be limited to LSR and plastic. LSR can be co-molded with other LSRs or TSEs offering dual hardness or two-color molded articles. Molding would be similar as described previously, but injection systems would need to suit the desired materials.

Due to the higher cost of tooling and processing development, LSR two-shot molding is typically considered for high-volume applications. Even as such, two-shot machines are inflexible and consigned to the slowest of the two materials' processes. Often, two separate machines may still offer less manufacturing cost and greater flexibility.

11.11   Conclusion

LSRs offer the potential for flash-free, low-pressure, quick-cycle molding processes that can create very cost-competitive products. LSRs can be an easy transition from plastic injection molding. LSRs may continue to encroach on the millable silicone molding sector, and are therefore an important advancement that custom molders should include as a core competency.

References

1. Toub, M., Silicone Elastomers, *Basic Rubber Technology*, Rubber Division of the American Chemical Society (2001), p. 498–514.

2. Fluid Automation, Inc., Wixom, MI.

3. Fluid Automation, Inc. Arburg open house, 2008.

4. Scheurell, A. M., Liquid Silicone Rubber Technology.

5. Engel, Guelph, Ontario.

6. Lange, C., New Development in Multi-component Injection Molding, International Silicone conference, Cleveland (April, 2003).

12 Secondary Operations and Additional Methods

TSE custom molded products offer a variety of designs and applications, ranging from general use products, such as O-rings, to highly engineered ultra clean products in the medical industry. These products may require specific and unique pre- and post-molding operations. Although it is not possible to touch on all methods, this chapter addresses major operational methods in generic terms. Some sections, particularly regarding coatings, are copied verbatim, or paraphrased from sources who supply equipment and/or services specific to the subject. Additional information is available by contacting these suppliers directly to suit specific needs.

12.1 Post Curing

Many TSEs have a cure profile that allows them to have sufficient state of cure to be removed from the mold without distortion, prior to achieving complete cross-linking. Articles can be subjected to additional heat as a secondary operation to complete the cure process. This post-curing process can take several hours at elevated temperature to satisfy physical properties. Post-curing drives off low molecular weight chemicals and reaction byproducts (outgas), expels residual cure chemicals, and oxidizes the outer surface of articles. Post-curing replaces the need to extend molding cure time, optimizing the molding operation. Generally speaking, post-curing is far less expensive than extending a molding cure time. Many molded articles can be post-cured in bulk, offering the advantage of high efficiency rates with little-to-no labor. Other, more delicate, articles may require fixtures or racks to keep articles from touching each other. Post-curing can cause indentations in the TSE, if articles are allowed to touch each other or are placed on surfaces that can leave an impression.

Most post-curing is done with hot air ovens. However, other methods such as UV and microwave can be used. Continuous ovens that utilize a type of auger can feed articles through an oven as a continuous process, as opposed to a batch process.

12.2 Material Filtering

Contaminants and/or undispersed ingredients can have an adverse affect on some end products molded from TSEs. Metal shavings, clumps of powders, or cured material can be lodged within the stock. These anomalies can be transferred into the molded article. Critical products for markets, such as medical or automotive braking, must be free of any of these contaminants. Thin-wall membranes may also require screening. Filtering material through screen packs is common for these industries. Screens can be purchased in a variety of hole sizes to meet specific end product requirements. Screening can be done as an extra operation such as extruding. For low viscosity materials, such as LSRs, filtering can be administered at the injection molding machine. In either case, screens need to be replaced frequently. In addition, screens can add shear stress to the material. Tests should be conducted to determine the amount of scorch introduced by screening.

12.3   Flash

As described in Chapter 1, TSE molding is susceptible to flash because of the slow curing process and the inherent thermal expansion of the TSE in the cavity during the molding process. Flash needs to be physically removed from the mold during the demolding cycle to avoid contamination in the next heating cycle. Also, mold vents need to be kept clean to avoid air entrapment. Mold vents quickly get plugged with flash, lubricants, adhesives, or cure byproducts. Ideally, mold parting lines should be physically brushed during every demolding cycle (even in processes considered flashless). This brushing, coupled with the difficulty of removing an elastic part from a mold (K.O. pins as used in plastic injection molding will quickly seize up with flash in a TSE molding process), are the predominant characteristics that prevent automation in a TSE molding process.

Overflow, or dump, grooves are frequently added to the mold along the periphery of the cavity at each parting line. Predominantly seen in compression molding, these grooves can also be incorporated in transfer and injection. Overflow grooves serve several purposes:

- Material fill variation — in compression molding, for instance, it is near impossible to predict exactly how much material to place in the mold to fill the cavity. Besides, prep shape variation and the inherent thermal expansion of the TSE generally require that an excess of material be introduced into the mold to assure complete fill.
- Flush air and contaminants — as mentioned previously, TSE molding has the propensity to outgas during the curing process. Therefore, it is best to flush TSE through the cavity and into an excess dump outside of the cavity. In addition, contaminants that are transported along the material's flow front are expelled in a non-critical overflow and discarded after molding.
- Maximum flash extension — overflow grooves need to be machined adjacent to the periphery of the cavity in the mold. The proximity should be something less than the maximum allowed flash extension for the molded part. The grooves need to have a geometry that creates a high stress concentration between the periphery of the molded article and the overflow groove. The stress concentration allows for a clean break when the overflow groove is removed subsequent to molding. Figure 12.1 shows three versions of overflow grooves that are machined adjacent to the cavity of the mold. Version A may offer a clean break from the cavity in terms of a stress riser, but makes the mold very fragile. Notice the thin vertical edge of steel on the inner side of the overflow groove; this edge is prone to damage by internal cavity pressure and excess flash left on the parting line. Version B minimized the thin vertical wall by angling away from the cavity. However, where possible and practical, version C should be used. This version has the overflow groove machined into the opposite half of the mold, mitigating mold damage from thinned sections.

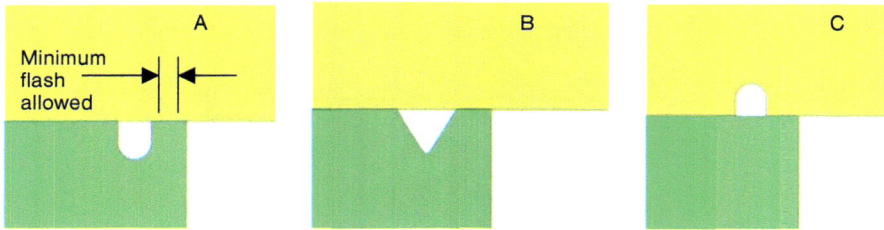

Figure 12.1 Overflow groove versions

12.3.1   Cryogenic Deflash

Flash is evident on a molded part at every parting line. Regardless of the mold construction (even those considered flashless), minute flash can be detected under magnification. A possible exception would be compression mold and die-cut. However, technically the pad is considered flash, it is simply thick enough to be considered part of the end product and will not dislodge. In some instances, any residual flash can be a detriment in the end use of the product. Consider a seal for a medical delivery device, or blood transfusion: if flash can dislodge from the molded seal and be transported in the flow of fluid, it can be fatal to the patient.

Cryogenic deflashing has long been used to remove thin flash from molded parts. The process incorporates a machine that is a large enclosed box and inside is a basket in which molded articles are placed. The machine is plumbed with liquid nitrogen which can lower the temperature inside the box down to approximately −240 °F. For a specific part to be cryogenically deflashed, the temperature should be set below the materials glass transition temperature. The idea of this machine is to make thin sections of the desired article brittle. Once at that brittle point, the machine tumbles the articles inside the basket, and the collision of articles against each other removes the brittle flash from the articles. A more aggressive deflashing process can incorporate high speed deflashing media particulate directed at the tumbling parts.

Cryogenic deflashing requires very careful attention to temperature, stabilizing time, tumble time, and media shot RPM (if used). Additionally, it is important to allow the machine to stabilize to its set temperature before processing parts. Exceeding the materials glass transition temperature for too long can result in fractured parts. Cryogenic deflashing is a matter of getting the thin flash to reach its glass transition temperature, using the correct impeller RPMs and, most importantly, choosing the appropriate length of cycle. Cycle time is extremely crucial so as not to over-deflash the parts. Therefore, time and temperature relationship is crucial. Too long in the machine and the finished parts are destroyed, but too little and the flash will not be removed. Cryogenic grade deflashing media should be used when trying to achieve short deflashing cycles.

Cryogenic deflashing has its place. However, avoiding deflashing altogether should be the goal in establishing a TSE process. Cryogenic deflashing is not only an added operation but it is another process that can create scrap. Damage by deflashing is common when the incorrect

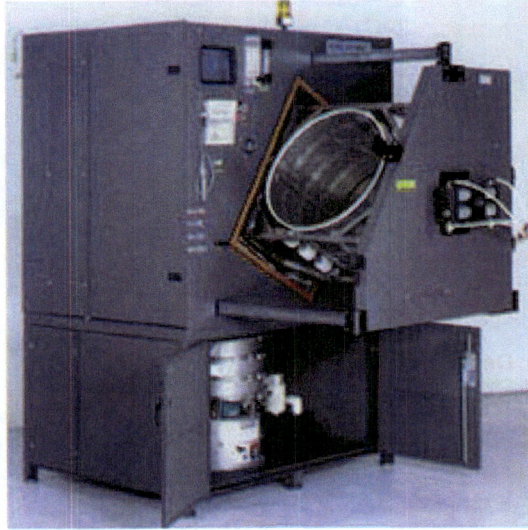

Figure 12.2 Cryogenic deflash unit (Courtesy of Cryogenic Deflash Systems) [1]

deflashing parameters are used, or the process is not stabilized. This finishing process can add additional inspection requirements to assure damage did not occur. Also, washing of parts may be required in some cases to remove potential flash particles that may get lodged in undercuts on the final article.

On the positive side, if carried out properly, cryogenic deflashing can be an acceptable process, especially on high quality/critical products where no flash remnant is allowed. Other flash removal methods, such as die cutting or knife trimming, may also require an additional 100 % inspection from cuts caused by knives and/or damage done by the use of other finishing methods.

12.4   Coatings

Coatings associated with the TSE industry cover a wide variety of materials and techniques. Coatings could be applied to substrates to be bonded during the vulcanization process. Coatings can also be applied to vulcanized TSEs. Before coatings are applied, surface treatment and cleanliness require careful attention on the substrate and the TSE. Coatings are used to promote adhesion, protect against oxidation, alter the surface friction, increase biocompatibility, and others.

12.4.1   Parylene Coating

Parylene coatings are used to protect products in electronics, aerospace, and medical device applications. Unlike most coatings that are applied in a liquid form, Parylene is a high-purity

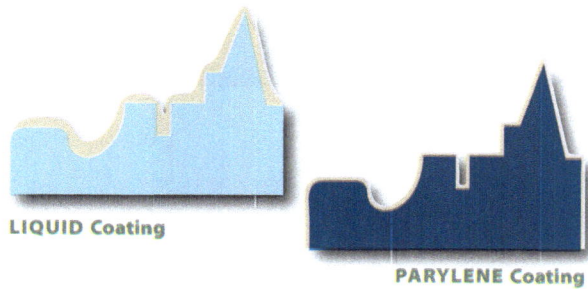

Figure 12.3 Comparison of liquid vs. Parylene coating (Courtesy of Para Tech Coatings, Inc.) [3]

powder known as a dimer. Parylene coating is thin and conformal, has no pinholes, and is resistant to organic solvents, inorganic reagents, and acids. The coating, when applied to an elastomer, is also uniquely flexible and will not crack or flake as the substrate is stretched and twisted. This coating serves several purposes, including electrical insulation, sterilizability, biocompatibility, moisture and chemical isolation, mechanical protection, enhanced lubricity, and surface consolidation to avert flaking or dusting [2].

Parylene can be applied to a number of materials and substrates. For TSE coatings, friction reduction and biocompatibility are most important. In medical applications, Parylene provides an inert barrier against moisture, chemicals, bio-fluids, and bio-gases. Parylenes N and C comply with the USP's (United States Pharmacopeia's) Class IV biological test requirements, which are necessary for long term implantables [3]. Parylene's static and dynamic coefficients of frictions range from 0.25 to 0.33, which allows coated TSE to approach the dry film lubricity of PTFE (Teflon®) [4].

Parylene coating is created by conversion of a crystalline dimer (a fine granular white powder) by heat to a gaseous form, then to a monomeric gas, and finally to a polymer. The powder is first vaporized at approximately 150 °C in a 1.0 torr vacuum, and the gas is heated to approximately 680 °C at 0.5 torr in a pyrolysis chamber to yield the monomeric diradical para-xylylene. The monomer gas enters an ambient temperature deposition chamber (at 0.1 torr vacuum), where it absorbs and polymerizes on the substrate [6].

12.4.2   Plasma Treatment

Plasma surface treatments offer ultra-clean surfaces, excellent preparation for promotion of adhesive bonding and printing, and functional group grafting for improved bio-compatibility. Plasma can be used to treat most substrates, including plastic (including PTFE), TSE, glass, and metals.

Energy will transition matter from a solid, to a liquid, and then a gas. As energy is introduced to the gas, some atoms begin to release their electrons, and the resulting electrically conductive gas becomes plasma. Plasma is often referred to as the fourth state of matter. Found in virtually every home and business, gas plasma remains a mystery to most. Common forms of plasmas include lightning, arc welding, neon signs, and fluorescent lighting [7].

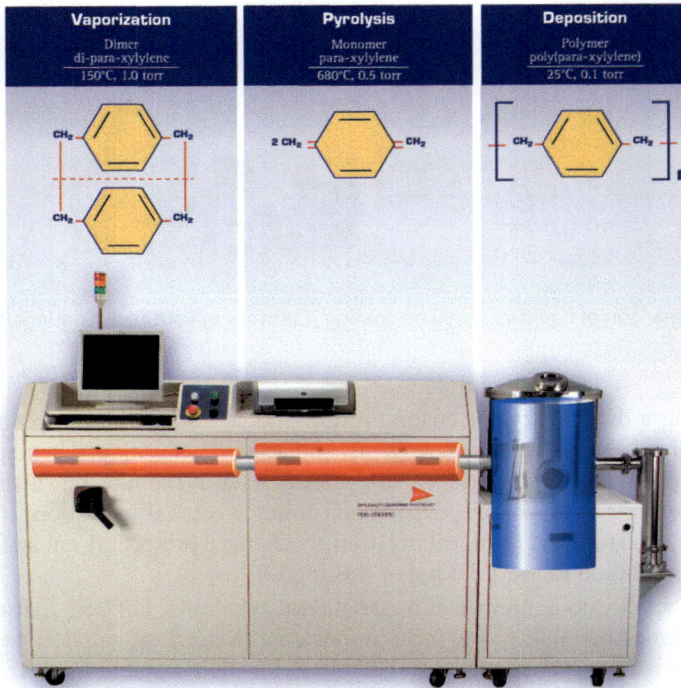

Figure 12.4 Parylene deposition process (Courtesy of Specialty Coating Systems) [5]

Common applications for plasma treatment include:

- Cleaning and removing organic films and contamination
- Etching and surface roughening
- Increasing adhesive bond strength
- Increasing or decreasing wetability
- Increasing biocompatibility
- Deposition of coatings
- Decreasing surface tack or reducing surface friction

Specific to TSE, plasma is mainly used to promote adhesion. Conceptually, TSE adhesion is difficult. TSE adhesion is performed during the molding/vulcanizing process, or post vulcanization, when rubber-to-rubber or rubber-to-other material bonds are required. Similarly, many post-vulcanization operations require printing or coating. In either event, cleanliness and surface wetability is essential for comprehensive adhesion.

Plasma removes a thin layer of organic contamination that is present on most surfaces which may be only a few monolayers thick. Plasma converts surface contaminants into small molecules, such as CO_2 and H_2O, which can easily be removed during the plasma process. In terms of wetability, plasma chemically alters the surface to receive polar liquids such as water. A hydrophilic surface results in more complete flow of inks and adhesives. Figure 12.5 illustrates increased wetability on polyurethane.

In addition to cleaning and wetting, plasma chemically modifies polymer surfaces by grafting functional groups onto the polymer backbone. Figure 12.6 shows XPS (X-ray photoelectron spectroscopy) analysis of the untreated surfaces of polypropylene. These data show dramatic changes in the surface chemistry caused by plasma treatment. The incorporation of oxygen functional groups into the surface of the polymer provides chemical bonding sites for the ink and adhesive.

Physical etching of a polymer surface is possible when subjected to direct plasma. This etching is sometimes referred to as micro-roughening. Figure 12.7a shows a pre-plasma RMS roughness of 1.87 nm, while Fig. 12.7b records an after-plasma RMS roughness of 34.1 nm. The surface area of the part was increased by ~95 % [7].

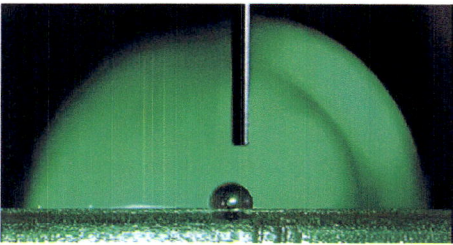

Figure 12.5a Before plasma treatment
(Courtesy of PVA TePla, America, Inc.) [7]

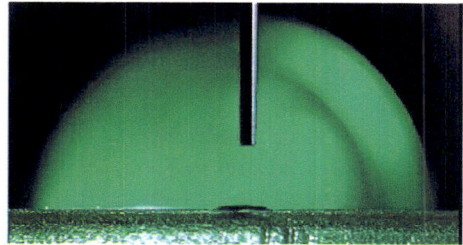

Figure 12.5b After plasma treatment

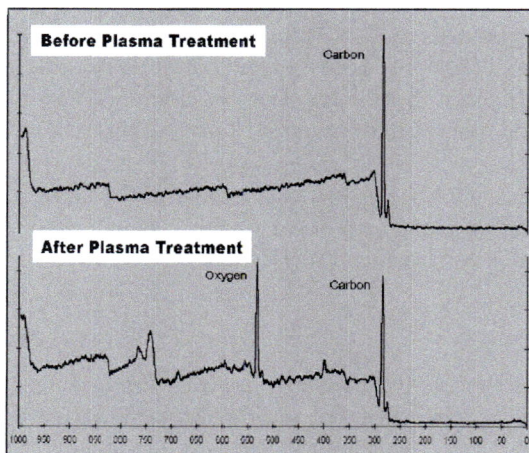

Figure 12.6 XPS analysis of polypropylene before and after plasma treatment (hydrogen is not detected by XPS (Courtesy of PVA TePla, America, Inc.) [7]

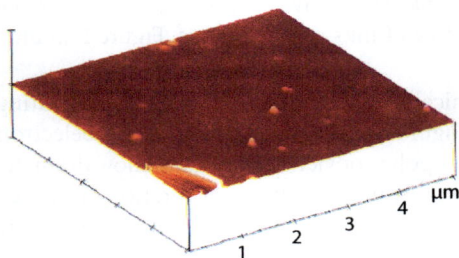

Figure 12.7a Before plasma treatment
(Courtesy of PVA TePla, America, Inc.) [7]

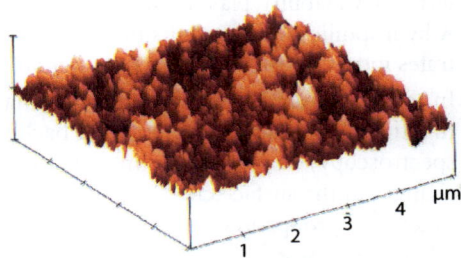

Figure 12.7b After plasma treatment

12.4.3   Chlorination

Surface chlorination of non-polar TSEs, such as NR, polyisoprene, polybutadiene and others, has long been used to promote post-molding adhesion and reduce surface friction. However, surface chlorination can also remove surface contamination and increase biocompatibility.

Chlorination of NR leads to thermodynamic, chemical, and mechanical changes to the NR surface. There are several methods used to chlorinate: exposure of the surface to chlorine gas; immersion in acidified sodium hypochlorite; and treatment with solutions of organic chlorinating agents, such as trichloroisocynauric acid (TCI). Chlorination actually alters the polymer backbone on the outer surface of the TSE. The thickness of penetration is on the order of 100 to 200 microns, depending on emersion time [8].

The surface chlorination process is an oxidation (bleaching) of the TSE surface. This process has been used successfully for years in the manufacturing of natural and synthetic rubber gloves to promote donning and prevent sticking [9].

Typical chlorination processes place TSE parts into a drum where chlorine gas is introduced. Parts are tumbled for a pre-specified period of time. The chlorine is then neutralized and rinsed thoroughly. Since there is an extensive cleaning to eliminate residual chlorine chemicals, powders and other surface contaminations are also removed, leaving a very clean surface [9].

Although surface chlorination for TSE is well known in the manufacturing of windshield wiper blades, gloves, galoshes, and baby bottle nipples, it has gained popularity in the medical devices industry where bio-burden and surface tack are a concern.

12.4.4   Oils

Silicone oils can be used to coat TSE parts to improve surface tack. Since silicone is very inert, it can be used in medical devices without adverse allergic affect to the patient. In fact, many syringe plungers are silicone coated even though medicines may come in contact with the silicone oil, which may be injected into the patient's blood stream. There are numerous methods of application and many silicone oils available to coat TSE parts. Silicone oils can be applied onto almost any polymer. Generally, parts are washed and cleaned and then siliconized. The silicone oil is usually a high-viscosity fluid, and is applied through an emulsion process.

Some TSEs are more porous than others and can easily absorb fluids. Silicone is one such example. Various oils can be added to an uncured silicone polymer past its saturation point, and then molded into the final product. Once the product cools down, the oil will start to leach out of the product. This condition is used successfully in self-lubricating (or self-leaching) applications, where assemblies may require less friction. For example, sealed electrical connectors for the automotive industry use self-lubricating silicone in the connector seals so that tiny wires can pass through the holes without threat of bending.

12.5   Adhesion

Having an elastomeric component attached to a rigid substrate makes for a desirable and versatile product. For example, radial shaft seals use various TSEs that are bonded to metal shells to provide a stiff outside diameter that may press-fit into a housing, and offer a flexible inner membrane that may seal to a shaft. In-mold bonding of TSEs to metallic and non-metallic substrates is a common practice in TSE molding.

There are essentially three elements required for adequate adhesion: the TSE, the adhesive, and the substrate. The base polymer and substrate are selected by the product designer to operate effectively in a given environment. Provided that the TSE can flow into the mold without developing a significant amount of crosslinking (less than 2 %), a bond can be formed with any TSE [10].

The choice of bonding agents relies primarily on the base polymer and its cure system. Other issues to consider are the TSE hardness and the substrate surface being bonded. Adhesive suppliers have charts and guidelines to suggest an appropriate adhesive that will work in a given condition. Some applications may require a two-coat system that incorporates a primer coat followed by a top coat. Primer/adhesives can be applied by dip, brush, or spray method.

The specific bonding ingredients are proprietary to adhesive suppliers and are usually covered under various patents. However, in generic terms, the bonding ingredients used are primarily dissolved thermoset elastomers, thermoset phenolics, organo-functional silanes/siloxanes, and other additives [11].

12.5.1   Dipping

Dipping substrates into a primer/adhesive solution may offer the fastest application method. The dip method requires a tub or vat that contains the primer/adhesive diluted with a solvent. Follow dipping instructions for the type of solvent and dilution as specified by the adhesive supplier. Experimentation should be done to determine the best bond by varying dilutions. Parts can be hung by pegs or hooks and dipped into the solution and allowed to air-dry.

Advantages of dipping:

- Fast
- Inexpensive capital investment

- Good for delicate or cumbersome parts
- Little-to-no maintenance
- Simple experimentation to launch into production
- Minimal procedures

Disadvantages of dipping:

- Heavy concentration of primer/adhesive buildup on the bottom side of hanging parts due to gravity
- Heavy labor effort
- Waste of adhesive materials
- Difficult to evacuate potentially harmful fumes
- Greater potential for contamination because of the amount of space required

12.5.2   Tumble Baskets

Tumble baskets incorporating a fixed sprayer may be the most efficient application of primer/adhesive. Tumble baskets, such as the ones shown in Figs. 12.8 and 12.9, can be completely enclosed with the capability of adjusting internal temperature. Temperature control is significant, especially when using environmentally friendly water-based adhesives. Many primer/adhesives distribute better when heat is applied. In addition, heat can be administered subsequent to spraying, while the basket is still tumbling, to reduce the amount of drying time.

Dual spray nozzles incorporated into a tumble basket offer the capability of applying primer and adhesive in the same operation. Careful attention should be paid to make sure the primer is adequately dried before applying the adhesive. In many cases, temperatures, times, and

Figure 12.8 Enclosed adhesive spray booth (left), Section view (right)
(Courtesy of Walther Trowal Co.) [12]

Figure 12.9a Tumble basket for spray booth
(Courtesy of Walther Trowal Co.) [12]

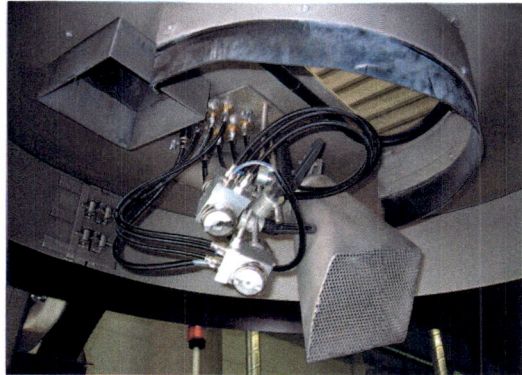

Figure 12.9b Spray gun inside booth

basket RPM may require different settings for primers vs. adhesives. PLC controlled machines are recommended for tumble basket spray applications. Consideration to the volume of parts in the basket and the part shape/weight is critical for adequate primer/adhesive distribution. Large or heavy parts or parts prone to entanglement may not be good candidates for tumble spraying.

Extensive development is required to establish the best process for adequate adhesion. DOEs (Design of Experiment) should be done using all, or some, of the following as variables: spray dilution, spray duration, basket RPM, volume of parts in a basket, temperature for spraying, temperature for drying, and/or drying time. If equipped with a PLC, and once the process is proven and established, these parameters can be downloaded and saved to avoid potential set-up errors.

Spray head technology has advanced over the years to prevent clogging. However, controls should be put in place to assure spraying is properly administered. Surprisingly, spray systems can be changed over to new adhesives in as little as 10 minutes.

Advantages of tumble spray:

- Even application of primer/adhesive
- Controlled environment
- Options for temperature during spraying and drying
- Better repeatability
- Low labor cost
- Easier evacuation of potentially harmful fumes
- Less floor space
- Less threat of contamination

Disadvantages of tumble spray:

- Relatively expensive capital investment
- Time consuming to establish best process (production launch)
- Potential part entanglement

12.5.3   Chain-On-Edge

In the event that a location-specific primer/adhesive application is required, a chain-on-edge machine may be a suitable option. This machine can also be used for special coatings or sealants. Many adhesives create mold fouling during the molding process. Minimizing the amount of adhesive on the substrate reduces the chance of mold fouling. Chain-on-edge works best for round parts. This method incorporates a series of equally spaced spindles that are laced and looped together with a chain and are slowly pulled through a track. The spindles have a nest mounted on the top to accept substrates to be sprayed. At the bottom of the spindle is a gear, which, when activated with a meshing gear, causes the spindle to turn. Toward the middle of the spindle is a bearing that is attached to the chain to allow the spindle to spin independently. This continuous series of spindles moves through various stations within the chain-on-edge machine.

Figure 12.10a Rotary table machine
(Courtesy of Turbo Spray Midwest) [13]

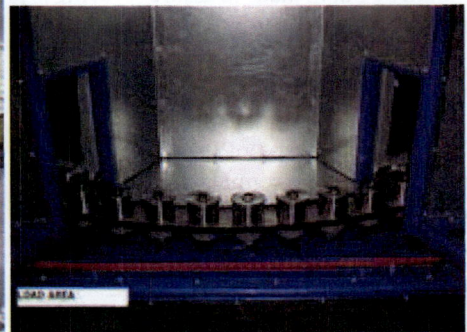

Figure 12.10b Close-up of table

The first station of the machine is a load station. Here the substrate is placed onto the nest either automatically or by hand. The next station is the spray application. As the spindle is moved along the track, the gear at the bottom of the spindle meshes with a turning gear (this can also be a friction drive) where it forces the spindle to turn. As it continues to spin, and is carried along the track, it passes by a spray nozzle that is strategically positioned to selectively apply primer/adhesive. The spray nozzle is usually only activated once the substrate is directly within the nozzle's path. Multiple nozzles can be used to apply in several targeted areas, or to apply primer and adhesive. Subsequent to the primer/adhesive application, the primer/adhesive needs to dry. Any number of methods can be incorporated to accomplish this, from air drying to microwave. The last station in the operation is the unload station. Here the substrates are removed from the chain-on-edge automatically or by hand.

12.5.4   Rotary Table

A rotary table is similar to a chain-on-edge machine, except that the spindles are fixed to a table. In this case, the table instead of a chain moves the spindles to various stations. On a rotary table spindles are generally aligned better than in chain-on-edge machines because bearings are fixed onto the table. For more accurate application of adhesive, rotary tables are recommended.

Advantages of chain-on-edge and rotary table sprayer:

• Selective application
• Minimizes mold fouling due to adhesive wash
• Even primer/adhesive distribution
• Good for delicate parts
• Good for secondary sealants and coatings that cannot contact the elastomer

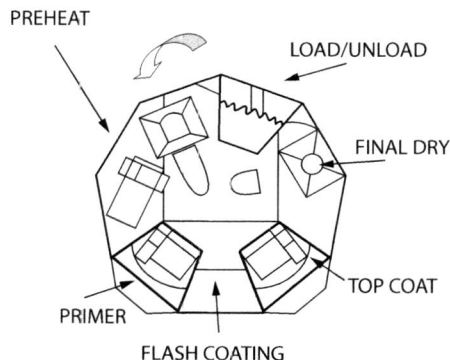

Figure 12.11 Rotary table layout (Courtesy of Turbo Spray Midwest) [13]

Disadvantage of chain-on-edge and rotary table sprayer:

- Expensive capital investment
- Slow process
- Maintenance intensive
- Requires large floor space
- Potential contamination because of large layout.

12.5.5   Other Application Methods

Masking, brushing, or lay-and-spray are other techniques for applying primer/adhesive to substrates. Many low volume or delicate parts require creative primer/adhesive application techniques.

Regardless of the application method, the instruction for solvents and dilutions from the supplier should only be used as a starting point. The primer/adhesive suppliers cannot predict the extent of potential applications and therefore make no claim to a one-size-fits-all approach. Experimentation should be done with different dilution ratios, application temperatures, and thicknesses. Heavy surface texture may absorb more adhesive than typically recommended by adhesive suppliers.

12.5.6   Self-Bonding Methods

Self-bonding ingredients can be added to compounds to enhance bond, or completely eliminate the need for primer/adhesive. These additives are generally covered under patents. Much of the science behind self-bonding is cloaked in secrecy. These ingredients can be targeted to specific substrates. The concept of self-bonding ingredients is a great idea. However, they do not work for every application. The self-bonding ingredients need to work in harmony with the compound's cure system and base polymer, and not interfere with other ingredients. Oils and fats may interfere with the self-bonding ingredient's ability to effectively bond. Furthermore, the self-bonding ingredients need to be compatible with the substrate to be bonded. Unfortunately, this is best determined through trial and error, and can be very time consuming.

Additional attention needs to be given to the molding operation. Since the very nature of these ingredients is to bond to substrates such as metal, the mold steel is susceptible to bond as well. Special platings, such as titanium-nitride or nickel-teflon, should be used on the molding surface of the mold. Spray mold releases that are applied prior to heating the mold and bake on to the molding surface work well. Spray mold releases may also be needed during regular production molding to prevent sticking. As with any type of bonding, the substrate to be bonded should be clean, dry, and free from oils or any ingredients that could adversely affect the cure or bond with the TSE.

Many LSR materials are now available as self-bonding. The self-bonding ingredients are premixed with the silicone. Since LSR is purchased pre-mixed, modifying the self-bonding ingredients or its concentration level is not an option. The same principles of self-bonding apply to LSR as they do to compounded TSEs.

12.5.7   Substrate Preparation

Almost any material can be bonded to TSE, provided it can withstand the TSEs molding environment. Chemical etching of polymers (even unlikely candidates such as PTFE, which are etched with sodium naphthenate) can be effective in preparing the substrate for adhesion. Bonding TSE to steel is a more traditional technique that requires either physically removing the oxidation layer from the steel by abrasion (grit blasting) or conversion coating with a phosphate process.

Grit blasting has several disadvantages. First, the velocity of grit hitting the metal insert can cause distortion. Second, the grit used must be carefully removed, or mold damage may occur if the grit is carried along to the molding operation. Third, the grit-blasted insert has a very limited shelf-life. Inserts need to be immediately coated to protect for oxidation.

Conversion coating of steel inserts by use of phosphate is very common. Zinc phosphate is the most common, but the more environmentally friendly iron phosphate is gaining popularity. Typical zinc phosphate applications produce coating weights of 1.5–3.5 $g \cdot m^{-3}$, and crystal size do not exceed 5 µm in diameter [14]. Phosphate coatings provide a clean, protected, and porous surface for adequate adhesion to TSEs. Additional corrosion protection can be added by a chromate or di-chromate application.

An efficient method for applying zinc-phosphate coating to metal parts is through a continuous batch process. This method carries parts through the required baths using a large diameter perforated drum with helix flutes. Figures 12.12 and 12.13 show a continuous batch drum phosphate and wash machine. Spray nozzles inside the drum saturate the parts being carried along in the barrel with fluid from their respective baths. These machines can also apply adhesives

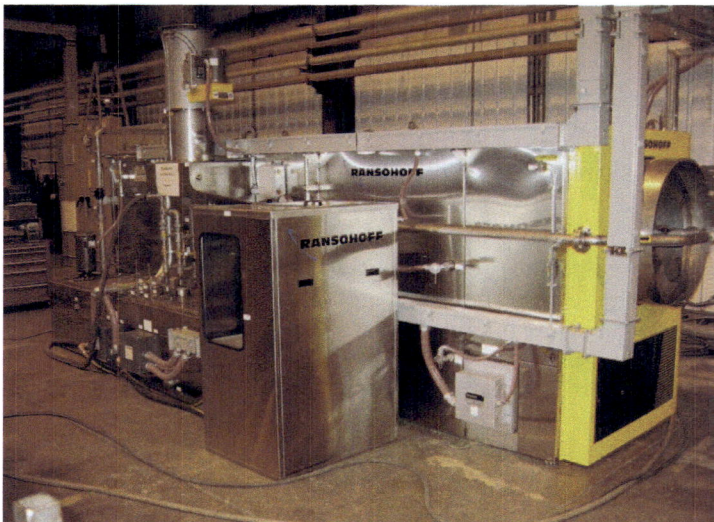

Figure 12.12 Continuous batch drum phosphate and wash machine (Courtesy of Ransohoff, a division of Cleaning Technologies Group LLC) [15]

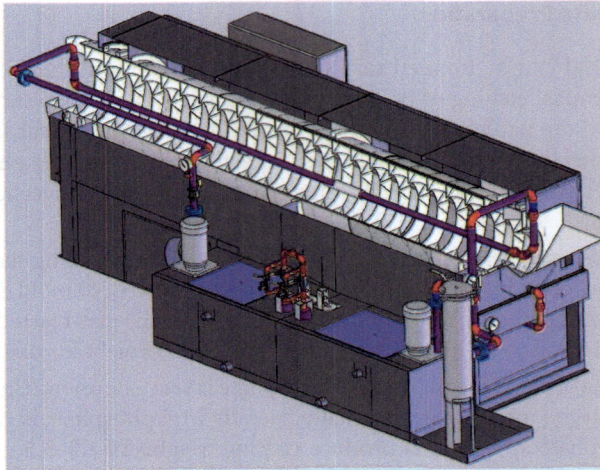

Figure 12.13 Cut-away view of continuous batch drum phosphate and wash machine (Courtesy of Ransohoff, a division of Cleaning Technologies Group LLC) [15]

or coatings after zinc-phosphate. The last station is usually a drying station, so parts are ready to be used once they exit the machine.

Anodizing is a common method for treating aluminum surfaces. Since anodizing leaves a porous surface, a sealant is usually applied to protect the anodized surface. However, sealants are not recommended, if the aluminum is intended to be bonded to a TSE. Even stainless steel can be bonded to TSE. Regardless of its connotation, stainless steel develops an oxidation layer over time. To be effectively bonded to TSE, this oxidation layer needs to be removed physically or through passivation before the application of adhesive.

12.6   Conclusion

Custom molding of TSEs can encompass a wide variety of pre- and post-operations. It is a good idea to become familiar with them because invariably the custom molder will be required to undertake some of these processes. The reader is encouraged to contact the references below to gain more information on specific processes.

References

1. Cryogenic Deflash Systems, Santa Anna, CA.
2. Para Tech Coating, Inc., *Parylene Technology* (2008), www.parylene.com.
3. Para Tech Coatings, Inc., Aliso Viejo, CA.
4. Humphrey, B., Transparent Film adds Value to Elastomers, special *to Rubber World* (March, 1999).
5. Specialty Coating Systems, Indianapolis, IN.

6. Olson, R., *Parylene, a Biostable Coating for Medical Applications,* Specialty Coatings Systems, Inc.

7. PVA TePla, America, Inc., Corona, CA.

8. Moore, M. J., Chlorination of NR: Surface and Failure Analyses in a Post-Vulcarization Bonded Composite, *Rubber World,* (November, 2001), volume 225, no. 2.

9. Amdur, S. Ph. D., Skariah, A. Silica enhanced Chlorination, Resistivity in Latex Gloves, *Rubber and Plastic News,* August 11, 2003.

10. Lindsay, J., Rubber to Metal Bonding, *Materials World* (May 1999), Vol. 7 no. 5, p. 267–268.

11. Cassel, R. and Kistner, D., Potential Solutions to Processing Clear Adhesives, *Rubber and Plastics News,* (October 6, 2008).

12. Walther Trowal Gmbh and Co. KG, Haan, Germany.

13. Turbo Spray Midwest, Suwanee, GA.

14. Cassel, R. and Kistner, D., Potential Solutions to Processing Clear Adhesives, *Rubber and Plastics News,* (October 6, 2008).

15. Ransohoff, a division of Cleaning Technologies Group, LLC, Cincinnati, OH. Contact Barney Bosse, V. P. of Engineering.

13   TSE Molding Processing

As described in previous chapters, TSE molding encompasses many different molding processes. This chapter will cover TSE molding as it relates to establishing and troubleshooting the process. Concentration will remain limited to custom TSE molding: injection, compression, and transfer molding, when describing the molding process. The following descriptions are general and can be incorporated in a variety of methods. However, since it is impossible to describe a process that will work in every environment, the reader should use discretion in interpreting the following process descriptions. Some adjustments for unique processes may be warranted.

Establishing a robust molding process requires a tremendous amount of planning, prototyping, and testing. Unlike plastic injection molding, TSE molders are best served if a cradle-to-grave approach is incorporated. This chapter recommends the steps required to cover a successful product launch.

13.1   Prototype

Building a prototype mold and molding sample parts is crucial for most TSE parts. The very nature of TSE suggests product designs that can be squeezed and sealed in-place, and/or offer flexible/stretchable members to seal and/or absorb dynamic motion. Therefore, product designs may include ribs, undercuts, lips, and/or protrusions that are required to flex or squeeze upon mold-open and/or part removal. Molds with these conditions would be considered unique to TSE, since any other moldable material would be destroyed by the mold movements.

Particularly in the event that mold opening and part removal can potentially damage the part, prototyping is imperative in the discovery process. However, other sound reasons to prototype include:

- Material shrinkage characteristics
- End product validation/testing
- Dimensional verification
- Parting line flash acceptability
- Cavity fill characteristics
- Air entrapment
- Mold opening and part removal sequence/limitations
- Cycle/cure time verifications
- Scrap-related issues
- Material-related issues

The prototype mold/process should emulate production intent. The prototype mold should be built identical to the production mold with the same production criteria:

- Steel type
- Mold hardness specification
- Plating/coating
- Tolerances and insert clearances
- Surface finishes
- Parting lines
- Sprue/gate size, location quantity

In some cases, multi-cavity prototype molds should be built to simulate cavity-to-cavity interaction. In the prototype phase, the compound is evaluated as well as all other aspects of the process. The chemist needs to be involved in the processing of prototypes and determine if processing aids and/or material enhancements need to be added to the TSE. Materials such as LSR are limited in terms of modification, but gums can be easily altered as needed. For obvious reasons, material changes are best done before the production phase.

The prototype phase is more than simply making a representative product. It is a learning tool aimed at eliminating potential problems down stream. Providing a stable process during prototyping will lead to a seamless transition into production. Therefore, it is crucial that the same molding shop and the same mold builder are used for prototyping and production. A formal submission plan including capability studies, full dimensional verification of parts and mold features, control plans and run-at-rates should be included, even if not specifically required by the customer.

For production, the fastest cycle time possible without producing defective parts is the desired outcome. This mindset needs to be the focus while prototyping. DOEs should be run, aimed at reducing the cure time as much as possible. Therefore, as the cure time is reduced, the mold temperature will need to be increased. Variations of time and temperature should be performed until undercures and molding defects occur — the lowest cure time and highest temperature need to be determined. Results may not be obvious until statistically significant trials are completed along with state-of-cure results. As a general rule, in the operating temperature range of 200–400 °F, for every increase of 18 °F, the cure time is cut in half [1].

TSE certifications need to accompany the material to be used. If a production-approved material is determined suitable, statistical data needs to be generated to show that it is consistent with what would be encountered in typical production environments. Careful consideration of maximum and minimum shelf-life requirements needs to take place. Some materials need to set for a period of time before they are considered useable, and out-of-date material can skew prototype assumptions. If a new material is developed, and has no long-term history with respect to physical properties data, it is important to keep a record of the actual properties exhibited by the material used during prototyping so that it can be duplicated in the production phase.

The prototype phase in TSE molding is typically a departure from what is performed in the plastics industry. Often, plastic prototypes are manufactured by specialty prototype shops, utilizing aluminum molds or stereo lithography to reduce time and cost. Production molds are often built by a production molder without experiencing the benefit of prototyping. In plastics, there are usually less questions about how the mold will be built and how the process will relate to the tool or material. Material shrinkage is given by material suppliers and part removal offers less of a challenge.

13.1.1 Prototype Plan

As to the success of any endeavor, a plan needs to be established with measurable deliverables. Prototyping a TSE process is no exception. Often, more emphasis is placed on producing a part than learning how to make it. Chapter 14 covers Manufacturing Process Planning, where five phases of Advanced Quality Planning embrace a team approach, a plan for each phase, and measurable deliverables.

13.2   Production

Establishing a production process will be less problematic if lessons are learned throughout the prototype phase, and proper procedures, checks, and sign-offs are followed as outlined in phases three and four in Chapter 14. Production tooling, equipment, materials, sequences, quality criteria, process sheets, operating instructions, and set-ups are all derived from the prototype phase and translated into a production format. This is only a starting point.

13.2.1   Cure Time/Temperature

Material cure curves that show the relationship between time and temperature should be provided by the materials lab along with an estimated cure time and temperature for the production mold (point A in Fig. 13.1). Records from the prototype mold should also be used as a guide for an optimum cure profile. Since temperature swings from mold-open times during prototyping cannot be easily duplicated in the production mold, prototype cure profiles may need adjustments. These charts are a guide — a starting point — to begin experimentation with the production mold for cure time and temperature.

The cure profile is established in a lab at ideal conditions without temperature variation. Therefore, the process engineer needs to determine the temperature variation throughout the molding process. As soon as a mold opens and mold plates separate, mold temperatures immediately start dropping. Mold plates cool quickly, particularly thin plates that are not in direct contact with heating platens. Obviously, mold-open time needs to be minimized and made consistent to offer a process in-control. The process engineer needs to verify the temperature fluctuations throughout the molding cycle by monitoring the extremes as soon as the mold opens, and just prior to closing the mold. This should be done after the mold is sampled and an established cycle time is derived.

Figure 13.2 shows a cure profile. Notice the temperature range. This range will be a combination of the mold open time influence and the inherent variation throughout the mold during a given molding cycle. If the cure time/temperature combination falls short of curing the

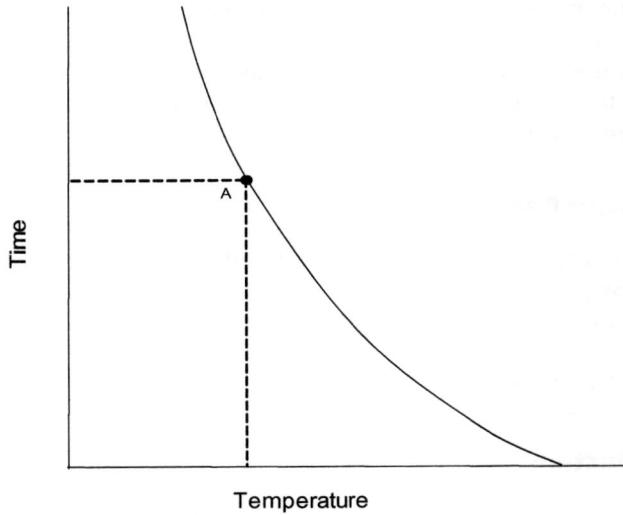

Figure 13.1 Time/temperature cure profile

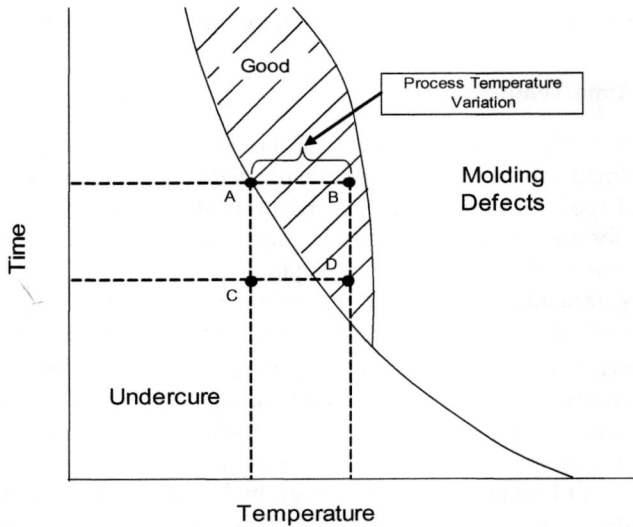

Figure 13.2 Time/temperature variation cure chart

molded part, it falls into the left side of the cure curve (designated as undercure). Conversely, too much cure time/temperature can also cause defects. A "sweet spot" is where ideal cure time/temperature combinations produce good products (designated as the cross-hatched area). From an economic standpoint, the faster the cycle time, the less expensive the end product becomes, and more parts are generated. The lowest temperature recorded throughout the established cycle will become the limiting factor on cure time — meaning, where the lowest temperature intersects the cure curve, establishes the cure time (point A). From this point, a corresponding cure time is provided on the chart and becomes the established production mold cure time. Notice that reducing the cure time would establish two new intersect points C and D. Although D is within the "Good" portion of the graph, C will produce an undercure. State-of-cure on the molded parts needs to be the ultimate determining factor for cure time/temperature.

13.2.2   Sampling Phase

The first molding cycle of a production mold is an anxious event. The first cycles should be performed by engineers and/or qualified technicians. This sampling phase is more of an exploratory approach to determine if everything is operating as anticipated. The following events should take place:

- Set-up press to estimated process
- Heat the mold to the desired set point
- Dry cycle the mold in the press and check for mold plate movements, binding, K.O. operation, and loading mechanism
- Verify cure time and cycle time
- Verify material's properties and shelf-life
- Run mold/press with estimated material use
- Check first run parts by cavity visually for any defects
- Fine-tune the process settings to produce good parts from each cavity (use trouble shooting guide referenced in troubleshooting chart below)
- Record all modifications and keep in the job folder
- Stabilize process — run minimum of 10 heat cycles consecutively

The second five heat cycles molded in the sampling phase should be considered stabilized, and should be given to QC to inspect per the inspection criteria sheets. The heat cycles will be analyzed for scrap and a pareto chart provided. If determined acceptable, several parts should be given to the chemical lab and analyzed for state-of-cure. If both studies prove favorable, all paperwork and instructions are updated, and the qualification phase is initiated. If either of the studies returns unfavorable results, the sampling phase is reinitiated and repeated until favorable results are achieved.

13.2.3   Qualification Phase

To understand the interactions of variation in key parameters in the molding cycle, a DOE needs to be run. The DOE will establish upper and lower control limits for the selected parameters. The DOE qualifies the process, not necessarily the mold. However, DOEs are very time-consuming. Therefore, a screening DOE is run first to set the stage for the full 3-factorial DOE. The screening DOE is a scaled-down version of a DOE that verifies the correct variables are considered in the full DOE.

A minimum eight hour run is performed to qualify the mold. For this run, the nominal parameters are set at the press. A trained production operator is required to perform this run. A minimum of 30 good pieces from each cavity are measured for critical characteristics (critical to the function of the part) and process characteristics (part feature which offers the potential for greatest variation — usually the largest dimension parallel and perpendicular to the mold parting line). From the 30-piece run, one part from each cavity is measured for every dimension on the print. The other purpose of the eight hour run is to determine the extent of mold fouling. This run will establish mold brushing and cleaning, and lubing (if allowed) frequency and will be recorded.

All capability study measurements need to meet a minimum 2.0 Cpk value. All DOE and dimensional verifications need to be within the upper and lower specification. If any of the above conditions fail, the following actions need to take place (either individually or in combination):

- Change the part print
- Revise the mold
- Change the material
- Revise the process

In the event conditions fail, the whole qualification needs to start over. If all conditions pass, the qualification package needs to be assembled and submitted.

13.2.4   Measurement Qualification

Measuring typical elastomeric parts can be challenging. The very nature of these parts would suggest deep undercuts and ribs making conventional measurement techniques ineffective.

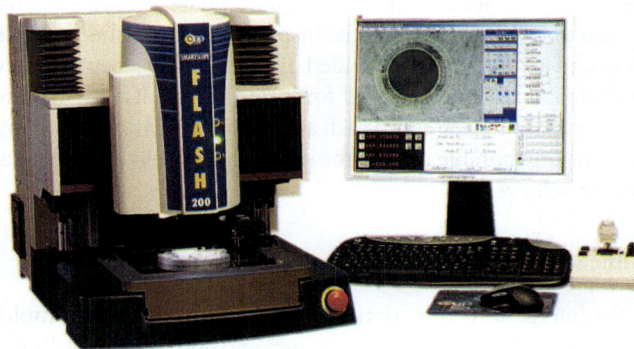

Figure 13.3 OGP non-contact measuring machine (Courtesy of Optical Gaging Products, Inc.) [2]

Non-contact measuring techniques should be used to qualify elastomer parts, because contact measuring machines will penetrate the substrate and can produce non-repeatable results. OGP (Optical Gaging Products) offers a variety of machines that use lasers or optics (non-contact).

Many TSE parts contain areas that cannot be measured without destructing the part. Additionally, many parts have built-in stresses, that when cut in half, will distort. For example, a radial shaft seal has several undercuts that cannot be measured without cutting the seal in half. However, the radial seal is also bonded to a metal shell. The elastomer shrinks when it is molded and exhibits built-in stress — cut in half the part will distort. A good technique for measuring elastomeric components is to cast and cross-section. This technique uses a pourable material that contains miniature glass beads (to eliminate shrinkage). As shown in Fig. 13.4, a molded part is placed in a container and the non-shrink material is poured over the seal. The casting is allowed to harden. The casting is then removed from the container and is cut in half as shown in Fig. 13.5. The cross-section can be placed on an OGP machine and the image can be enlarged to view undercuts and ribs. Once viewed optically in a cross-section, all hidden undercuts and ribs can be accurately measured. Moreover, the stresses created from the molding process are locked in place by the hardened casting, allowing for measurement that simulates a full-round part.

Many parts will require gages to submit the part to be measured consistently. Thought needs to be put into consistency in measurement. Gage R & R needs to be done on each measuring machine.

Figure 13.4a Seal placed in container

Figure 13.4b Non-shrink material being poured

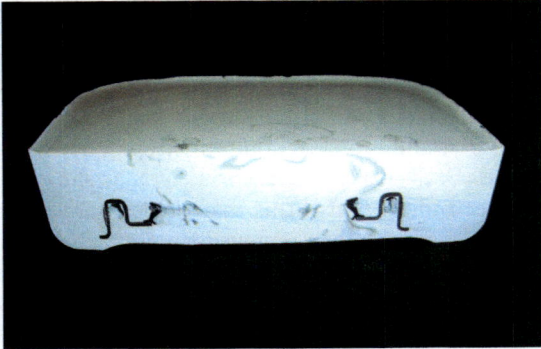

Figure 13.5a Casting cut in half

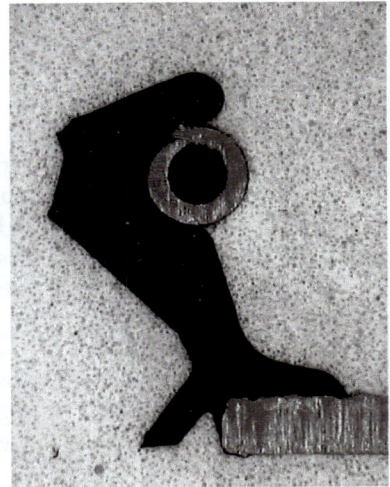

Figure 13.5b Close-up of cut casting

13.2.5 Continuous Improvement Phase

To this point, the qualification phase has centered on qualifying the mold, process, and equipment. The potential variation in the TSE material has been left out of the qualification. The material simply needed to comply with specific upper and lower specifications for certain physical properties. Considering that the TSE material variation is typically the largest contributor to molding variation, ideally the DOE should include key properties of the TSE. Unfortunately, it is highly impractical, if not impossible, to mix production batches to exhibit perfect line-to-line emulations of upper and lower control limits of specific parameters. Typically, the control limits are established from historical data, if an existing material is used, or arbitrary limits are established from similar materials or past experience by the chemist, if a new material is developed.

The qualification of material may sound subjective, but the validation does not end once the qualification package is complete and delivered. The first several production runs are monitored closely for scrap, cycle time, dimensional stability, and state of cure. The material's natural batch-to-batch variation in physical properties is recorded and overlaid onto the performance of the production molding run. If certain batches of material can be isolated to poor molding performance, the material's control limits are tightened accordingly. A good production manager and/or process engineer will evaluate the material periodically, if not every run. A long-term DOE should be performed, where incoming material displays maximum and minimum physical properties. This may take several weeks to find specific maximum and minimum values. A DOE should run sequentially. However, in the case of finding natural material variation, DOE sequence may need to be compromised.

Most modern injection molding machines provide data collection and SPC capabilities on all temperature, pressure and time parameters. Data is electronically collected over time and

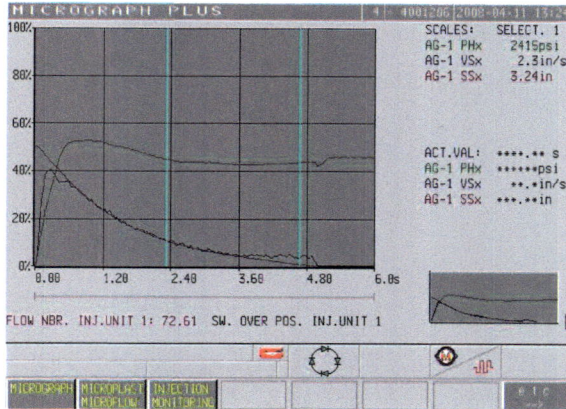

Figure 13.6 Injection profile graph (Courtesy of Engel) [3]

can statistically determine upper and lower control limits over historical runs. Alarms need to be triggered when control limits are breeched on critical parameters, warning the operators that something is out of control and needs to be rectified.

13.3 Troubleshooting

Utilizing a well thought-out qualification plan, and following the established process will avoid critical problems in production. Following the established process and administering continuous improvement is a challenge in a production environment, particularly in a growing company where there is a significant demand on resources.

Figure 13.7 shows the typical life of a production process including various stages. As greater emphasis is placed on success from all departments, the profitability generally goes up. As other priorities take over, the process takes on a life of its own and generally profitability diminishes. This should come as no revelation and is not unique to TSE molding. Management must recognize when a trend of profitability is headed in the wrong direction and correct this trend. Six Sigma and Lean Manufacturing techniques should be implemented to bring the product back to a manufacturing standard.

Unanticipated problems will be encountered in production from time to time. For this reason, a troubleshooting guide should be used to help isolate and identify the problem. The first thing any production lead or process engineer must do, if a molding problem is encountered, is to review the process and see if it falls within the original process specification. After all, it was running good at one time. This sounds simple, but given the multitude of parameters and the potential negative interactions, finding the problem takes a tenacious disposition. Everything is up for challenge: from material shelf-life to any process parameter. Fishbone diagrams are useful in finding problems when divided into categories of people, materials, and equipment. A clear definition of the problem will enable a fine-tuned approach to finding the

Process life timeline

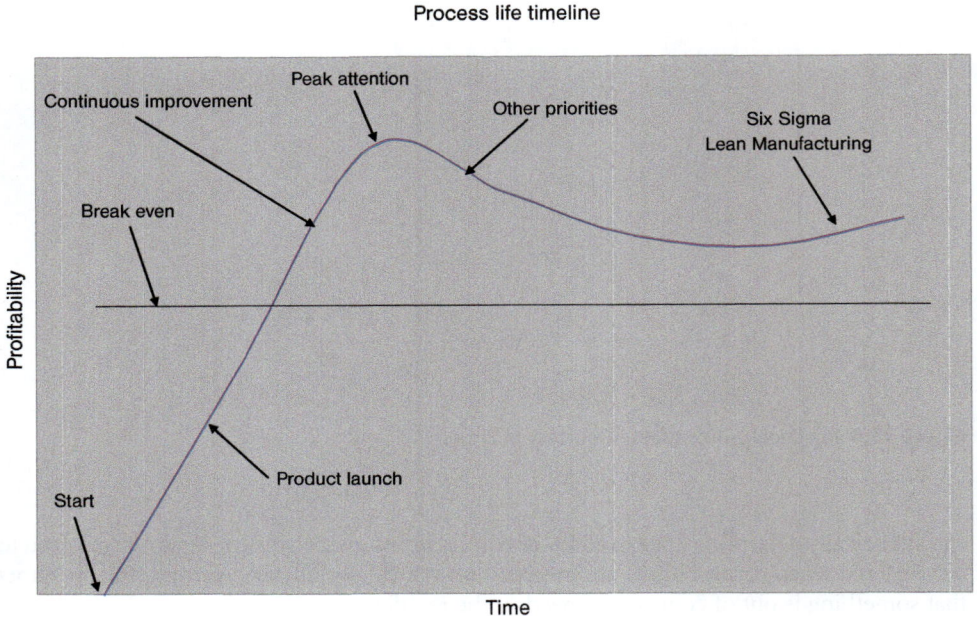

Figure 13.7 Product profitability over time

Figure 13.8 Backrind

Figure 13.9 Contamination

root cause. If nothing is unearthed, the troubleshooting guide as shown in Table 13.1 should be used.

As described previously, most modern injection molding machines offer SPC and data collection. This can be a useful tool in determining the root cause of problems. Temperatures, pressures, speeds, and times can all be retrieved from historical runs. Trends and spikes plot-

Table 13.1 Molding Process Troubleshooting Guide

	Blister	Non-Fill	Trapped Air	Sponge	Excess Flash	Knit Lines	Backrind	Contamination	Tear	Scorch	High Sprue/Gate	Sprue/Gate Chunk-Out	Non-Bond	Reversion	Under Cure
Increase Mold Temperature	X			X								X			X
Decrease Mold Temperature		X	X		X	X	X		X	X	X			X	
Add Bump	X	X	X	X		X	X								
Check/Add Vacuum		X	X	X	X	X									
Reduce Mold Lube	X		X	X		X									
Add/Revise Mold Lube		X				X				X					
Check/Increase Mold Venting	X	X	X			X									
Reduce Clamp Tonage	X	X	X			X	X								
Increase Clamp Tonage					X										
Increase Rubber and/or Gate Size		X				X					X				
Check Press/Mold Parallelism					X										
Plate Mold With Chrome, Or.						X			X						
Decrease Injection Hold Time					X		X				X				
Increase Injection Time	X	X				X									
Decrease Injection Time					X					X	X				
Increase Injection Chamber Temp						X									
Decrease Injection Chamber Temp					X		X			X	X			X	
Increase Injection Pressure		X				X									
Decrease Injection Pressure	X				X					X	X				
Increase Injection Speed						X									
Decrease Injection Speed	X	X	X		X		X			X	X				
Increase Stock Viscosity	X			X	X	X									
Decrease Stock Viscosity		X				X				X					
Check Moisture in Stock	X														
Check Dispirsion in Stock								X	X						
Check Stock Rheology		X	X	X		X				X					
Check Stock Shelf-Life	X	X	X	X		X				X	X				

ted over a period of time can indicate if the process was changed and by whom. It can also determine material flow problems. For instance, if injection speed is the setting on the press, the injection pressure chart demonstrates how hard the injection unit needs to work to push the TSE at a given speed. This data can be analyzed to see if the material's viscosity has changed.

Figure 13.10 Delamination

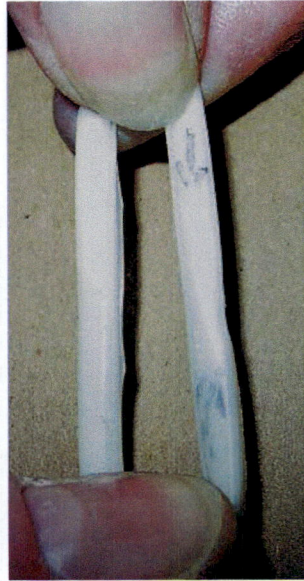

Figure 13.11 Undercure

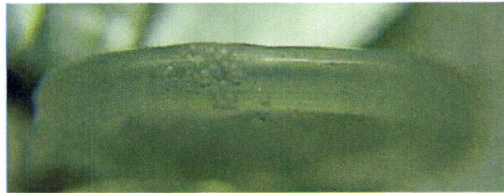

Figure 13.12 Sponge

Figure 13.13 Non-fill

Figure 13.14 High sprue

Figure 13.15 Non-bond

Figure 13.16 Knit line

Figure 13.17 Heavy flash

13.4   Conclusion

This chapter described the importance of pre-planning, and a cradle-to-grave approach. This system should be a good general guide for custom molders. Each custom molder should incorporate product launch and troubleshooting guidelines/procedures that are specific to their unique manufacturing process. Using these techniques will unlock the mystery of TSE custom molding.

References

1. Garvey, B. S. Jr., Accelerators of Vulcanization, *Introduction to Rubber Technology,* p. 117.

2. Optical Gaging Products, Inc., Rochester, NY.

3. Engel, Guelph, Ontario.

14   Manufacturing Process Planning

Manufacturing Process Planning (MPP) incorporates five linked phases of manufacturing development. MPP is not designed to simply comply to QS or ISO standards — although it certainly can be used for that purpose — it is designed to involve all departments that have a part in bringing a product into production to achieve buy-in, or to be made aware of potential new business, and to offer recommendations. Membership to the MPP process should include manufacturing, quality, chemical lab, tool engineering, process engineering, and product engineering. These members will be further referred to as "The Team". For MPP to work effectively, no phase can be started until the prior phase is brought to closure. Each phase must be documented and retained in the product master folder. It should be understood that the MPP process eliminates "majority rules" or dictatorial process decisions and instead relies on a series of events and checklists that need to be completed before proceeding to the next phase. MPP protects manufacturing from being forced into an approach without assessing the risks or qualifying its merits.

Phase I

This is the inquiry phase. A customer will send a request for quote to a manufacturing plant and request that the plant submit a formal quote, request for further information, or a refusal to bid. An MPP meeting is set up to determine whether the part should be quoted. If proceeding with a quote, the team must sign their portion of the Phase I form showing their acceptance of the quoted methods.

Phase I deliverables:

- Signed Phase I form (accept or reject) by team members
- Internal quote worksheet
- Letter to customer: quote, no-quote, or need more information
- Design/tolerance exceptions/recommendations

Phase II

Once a purchase order is received for prototype samples, Phase II is initiated. Preferably a design for manufacturing meeting should have taken place prior to Phase II. In the event that the design is already completed, a design review with the product engineers (in some cases the customer) needs to take place prior to initiating tooling/materials. Phase II is designed to not only provide prototype samples to the customer, but to validate the production intent process. A meeting is held with the MPP team members to kickoff Phase II.

Phase II deliverables:

- Signed Phase II form by team members
- Review of quoted process
- Kickoff of tooling, materials, and equipment
- Program timeline
- Prototype qualification plan
- Review of critical characteristics
- Review of quality acceptance criteria
- Prototype control plan

Phase III

Phase III determines whether the prototype portion of MPP met requirements and samples are suitable for shipment to the customer. In Phase III, parts are made with tooling, materials, and processes which best emulate the production intent. Phase III meetings cover product dimensional results and capability studies, as well as run-data, scrap rate, cure time, cycle time, material use, mold temperatures, state of cure, defect review, and any mold/process related concerns. Any discrepancies need to be rectified before closing Phase III.

Phase III deliverables:

- Signed Phase III form by team members
- Review of quoted process
- Full dimensional results
- Capability studies
- State of cure results
- Molding process sheets and set up sheets
- Run sheets (cure time, cycle time, temperatures, scrap percentage)
- Additional modification/evaluation recommendations
- Lessons learned report
- Assessment of meeting goals
- Customer submission package

Phase IV

The receipt of the production order kicks off Phase IV. A meeting is set up with the team to release production tooling, equipment, materials, packaging, etc. The qualification plan is reviewed. Scheduling and timing are reviewed. Goals for the production run are established. Production process sheets, routers, bill of material, packaging instructions, and work instructions are all created. Tool certifications, material certifications, and mold temperature mapping are created. The production process is trialed by engineering and all documentation is double checked. Once the process is established, the process is released to production operators to perform a minimum run trial. In this run, run-at-rate is established with actual material usage, and scrap rate. DOEs and capability studies are run as well.

Phase IV deliverables:

- Signed Phase IV form by team members
- Quote review
- Full dimensional results
- Capability studies
- State of cure results
- Molding process sheets and set up sheets
- Run sheets (cure time, cycle time, temperatures, scrap percent)
- DOE results
- Assessment of meeting goals
- Customer submission package
- Any information needed to be conveyed to the customer
- Any open issues

Phase V

This phase of the MPP is the follow-up phase to see if the production launch was a success, and if not, what needs to be done to make it a success. After several production runs, scrap, cycle time, material usage, and set-ups are all reviewed. Until Phase V is closed out, the production process is under engineering care. Meaning, production runs the process, but engineering is closely monitoring and continually improving certain aspects that are within the scope of allowable parameter change. Documents are updated as needed.

Phase V deliverables:

- Signed Phase V form by team members
- Quote-to-actual review
- Safety considerations
- Review of goals
- Lessons learned

Conclusion

Adapting this concept of MPP will not guarantee error-free production launches, but it will lessen the potential for unforeseen circumstances once a process reaches production. Incorporating a team approach to product launch offers a variety of opinions and draws on past experiences from a variety of factions. No individual person has all of the answers, particularly when a complicated process, such as TSE molding, is concerned.

Appendix 1: TSE Common Terms and Definitions

Accelerator, an ingredient in TSE compounds to increase the rate of crosslinking.

Air check, see trapped air

Backrind, torn, or jagged edge at the parting line of a molded article usually caused by thermal expansion.

Blister, a pocket usually caused by the expansion of trapped air.

Bloom, a change in the appearance of the surface of a molded article caused by the migration of a liquid of solid to the surface.

Bump, a short relief of clamp pressure after a press is closed to allow gas to escape from within the mold cavity.

Catalyst, a chemical that in small quantitiy activates a chemical reaction.

Cavity, the hollowed out area of mold that will fill with TSE to form the molded article.

Checking, short shallow cracks formed on the surface of a molded part.

Chlorination, a surface treatment process that oxidizes (bleaches) a TSE surface.

Cold runner, an insulted delivery system for injection molding that keeps the mold runners cool to prevent TSE from crosslinking.

Compound, a mix of material ingredients to produce a material that will be used for molding that will meet certain specifications and processing conditions.

Compounding, Mixing material ingredients to meet certain requirements.

Compression mold, a mold that uses the closing action of the mold to squeeze and displace an uncured perform to fill a mold cavity.

Compression set, the deformation retained in a part after a period of time under compression.

Crosslink, a chemical reaction that links polymer chains generally through heat and pressure. Also called cure, vulcanize.

Cryogenic deflash, the freezing of a molded article below its glass transition temperature to break off flash from the molded article. Generally parts are tumbled.

Cushion, the volume of TSE left in the injection barrel that did not get purged out during the injection cycle. Prevents the plunger from bottoming out in the barrel.

Cure, see crosslink

Dilatant, as shear increases in a fluid, viscosity increases. These fluids are also called shear-thickening fluids.

Disk spring, a high-pressure spring that is formed into a washer and has a cone shape. As the washer is flattened, the washer exerts a pressure to resist deformation. Used to keep cavity stacks closed when complete mold closure is in question.

Dump groove, see overflow

Durometer, a scale for measurement of hardness for elastomers.

Dumbell, a shape that is die-cut out of thin test slab that is used for many standard physical test.

Elasticity, A perfectly elastic material will conform to a deformation under force and have 100 % recovery once the force is removed.

Elastomer, a polymer that can be deformed significantly by a force and have an almost perfect recovery once the force is removed. Term is used interchangeably with rubber.

Elongation, an extension test produced by a tensile test.

Filler, an ingredient added to a TSE that could affect physical properties and processability, or simply reduce the cost of the compound.

Flash, a thin film of cured TSE that forms at mold parting lines as a result of material bleeding across the mold parting line.

Flash groove, an intentional groove cut into a mold to allow access material to fill into. Also used as a stress riser for trimming flash to a minimum extension.

Flashless transfer mold, a molding method that uses individual inserts stacks that use rubber pressure to keep cavities shut. Cavity stacks are usually vented to allow air to escape but not TSE.

Formulation, a mix of ingredients to a prescribed amount and procedure.

Flow marks, surface imperfections that are evident in a molded part that are caused by directional flow patterns during the molding operation.

Fouling, a byproduct of the molding process that leaves an undesirable residue in the mold.

Gate, a recess cut into the mold within the path of the uncured material flow that fills the cavity area.

Glass transition point, a temperature at which a material loses its flexibility

Green strength, the relative strength of the uncured material as it relates to premolding handling.

Gum, an uncured mix that has doe-like consistency

Injection molding, a closed-mold filling process that uses high pressure to force TSE through a series of runners and sprues to fill a cavity.

Insert, a term used for a carrier that is placed into a TSE mold and chemically or mechanically attached to the TSE through the molding operations. This term is also used to identify a mold cavity stack where separate interchangeable cavities are manufactured and "inserted" into holding plates.

Knit line, a demarcation in the molded article where visibly two flow fronts meet.

Land, typically a mold surface area adjacent to the cavity where to mold parting lines meet.

Mill, a machine that has two opposing turning rollers at different speeds that can masticate compound.

Modulus, the ratio of stress to strain

Monomer, the single chemical structure that is the repeating element of a polymer.

Mooney scorch, a measure of a materials time until crosslinking occurs.

Mooney viscosity, a measure of a compound or base polymers viscosity as measured on a Mooney viscometer.

Newtonian fluids, these fluids are considered ideal liquids and truly viscous. As shear rate changes the viscosity remains the same for these fluids.

Non-Newtonian fluids, these fluids are affected by shear.

Overflow, a machined groove around the periphery of the cavity to allow access material to flow into. Offers minimal flash extension. Also called tear bead, dump groove.

Parylene, is a high purity powder known as a dimer. Parylene coating is thin and conformal, has no pinholes, and is resistant to organic solvents, inorganic reagents and acids.

Parting line, the plane or surface in a mold that meets another plane or surface to form the closure of the mold.

Phosphate coatings, provide a clean, protected, and porous surface of steel carriers for adequate adhesion to TSE's.

Plastic, a remeltable hard polymer—see thermoplastic.

Plasticity, a material that exhibits no recovery when deformed by shear stress.

Plasticizer, an ingredient added to a polymer to soften or make more workable.

Polymer, a substance with a molecular structure linked end-to-end with predominantly identical molecules that repeat thousands of times.

Polymerization, subjects the monomer to an energy and/or chemical catalyst to attach predominantly identical monomers end-to-end to create a polymer.

Post bake, see post cure

Post cure, a process of subjecting a molded article to additional crosslinking generally done in hot air ovens to completely cure the article and/or drive off volatiles.

Plunger, the male portion (or piston) of a mold that closely fits inside of a pot for a transfer molding system, whereas the closure forces material into the cavity through sprues.

Pot, the female portion of a mold that mates with the plunger—see plunger.

Preform, an uncured geometry, generally preweighed, that is placed into a mold that ultimately gets formed and cured into the final article.

Pseudoplasic, as shear increases in a fluid, viscosity decreases. These fluids are also called shear-thinning fluids.

Retarder, an ingredient added to a TSE to slow down the onset of crosslinking.

Reversion, a symptom of overcuring—a deterioration of physical properties.

Rheology, is the study of change in form and flow of a material in terms of elasticity, viscosity, and plasticity

Rheopectic, Viscosity increases as a function of time. True rheopectic fluids are rare.

Rind, see flash

Rubber, see elastomer

Scorch, premature crosslinking given a molding process

Shear rate, is defined as the velocity over the given cross section with which molten or fluid layers are gliding along each other.

Shear stress, is defined as a stress applied parallel to a face of material.

Shear thinning, is the affect of a decrease in measured viscosity as shear rate increases, which is pseudoplasic or thixotropic behavior.

Shelf life, the useful life of an uncured material.

Sprue, a fill point for material into the cavity of a closed mold. Typically a small round hole that tapers away from the cavity to provide mass for removal at every cycle.

State of cure, a measure of the amount of cure that took place in a given molding process.

Surface finish, the finish or texture of a molded article and/or the mold.

Tensile strength, the maximum tensile strength applied to a specimen during stretching prior to rupture.

Transfer mold, a closed mold filling process by which a plunger forces uncured material through a pot and ultimately into the mold.

Trapped air, a surface anomaly where the TSE cured in the mold, but did not allow an air pocket to escape.

TSE, thermoset elastomer

Thermoplastic, a polymer that can be melted, solidified, and remelted and resolidified. Commonly just called plastic.

Thermoset, a polymer that irreversibly (or with difficulty) chemically crosslinks

Thixotropic, Viscosity decreases over time given a constant shear rate. As shear decreases, the material gradually recovers to the original properties.

Viscosity, is the internal friction of a fluid, caused by molecular attraction, which makes it resist a tendency to flow.

Vulcanize, see crosslink

Wasteless transfer mold, a transfer molding process that insulates the pot to a temperature below the TSE's crosslinking threshold.

Subject Index